A Gentle Introduction to Statistics Using SAS® Studio in the Cloud

Ron Cody

sas.com/books

The correct bibliographic citation for this manual is as follows: Cody, Ron. 2021. *A Gentle Introduction to Statistics Using SAS® Studio in the Cloud*. Cary, NC: SAS Institute Inc.

A Gentle Introduction to Statistics Using SAS® Studio in the Cloud

978-1-954844-49-0 (Hardcover)
978-1-954844-45-2 (Paperback)
978-1-954844-46-9 (Web PDF)
978-1-954844-47-6 (EPUB)
978-1-954844-48-3 (Kindle)

SAS Institute Inc., SAS Campus Drive, Cary, NC 27513-2414

April 2021

About this Book

What Does This Book Cover?

This book is designed to fulfill two purposes: one is to teach statistical concepts and the other is to show you how to perform statistical analysis using SAS Studio.

The book starts out with two short, introductory chapters describing statistical analysis (both descriptive and inferential) and experimental design. Following the introductory chapters are several chapters that show you how to register for SAS OnDemand for Academics, use some of the built-in SAS Studio tasks, how to upload data from your local computer to the SAS cloud, and how to convert data from multiple sources (such as Excel or CSV files) and create SAS data sets. There is one chapter on descriptive statistics, summarizing data both in table and graphical form. The remainder of the book describes most of the statistical tests that you will need for an introductory course in statistics.

Is This Book for You?

As the title suggests, this book is intended for someone with little or no knowledge of statistics and SAS, but it is also useful for someone with more statistical expertise who might not be familiar with SAS. One of the important points for beginners or people with more extensive knowledge of statistics, is a discussion of the assumptions that need to be satisfied for a particular statistical test to be valid. That is especially important because with SAS Studio tasks, anyone can click a mouse and perform very advanced statistical tests.

What Should You Know about the Examples?

Because you can download all of the programs and data sets used in this book from the SAS website, you can run any or all of the programs yourself to better understand how perform them.

Example Code and Data

You can access the example code and data for this book by linking to its author page at https://support.sas.com/cody.

SAS OnDemand for Academics

This book is compatible with SAS Studio and the SAS product called OnDemand for Academics. This is a cloud-based application that is free for anyone wanting to learn how to use SAS, not just college students. Although all the examples in the book were run using SAS OnDemand for Academics, you can run these tasks and programs on other versions of SAS Studio.

Where Are the Exercise Solutions?

Solutions to all the odd-numbered exercises are included at the end of the book. For those individuals who are not students, are working on their own, or are faculty members, please contact SAS Press for solutions to all of the exercises.

We Want to Hear from You

SAS Press books are written *by* SAS Users *for* SAS Users. We welcome your participation in their development and your feedback on SAS Press books that you are using. Please visit sas.com/books to do the following:

- Sign up to review a book
- Recommend a topic
- Request information on how to become a SAS Press author
- Provide feedback on a book

About the Author

Ron Cody, EdD, is a retired professor from the Rutgers Robert Wood Johnson Medical School who now works as a national instructor for SAS and as an author of books on SAS and statistics. A SAS user since 1977, Ron's extensive knowledge and innovative style have made him a popular presenter at local, regional, and national SAS conferences. He has authored or co-authored numerous books, as well as countless articles in medical and scientific journals.

Learn more about this author by visiting his author page at http://support.sas.com/cody. There you can download free book excerpts, access example code and data, read the latest reviews, get updates, and more.

Acknowledgments

I realize not too many people take the time to read the acknowledgments page. I wish you would because this book was really a team effort, and I would like all the gifted and talented people who worked on the book to get the recognition that they deserve.

I would like to start by thanking Sian Roberts who took on two roles. Not only did she take on the role of Developmental Editor for the book, even more importantly, she did an amazing job in getting this book published using a software application that none of us had used before. It was a struggle, and it required several video meetings to solve some sticky problems. Sian referred to me as the "Guinea pig" in trying to work with the software, but we later agreed that "beta tester" sounded better!

I normally thank Robert Harris last because his expertise comes after a book is mostly finished. I'm moving him up because he was also instrumental in figuring out the new publishing software. In his role as SAS Press artist, Robert created four cover designs and they were all spectacular, making our cover choice very difficult. With this cover he managed to give the images of "Gentle" and "Cloud."

This book is based on an older book called *A Gentle Introduction to Statistics Using SAS Studio*. That book used SAS University Edition while this book uses a free SAS product called OnDemand for Academics. Although we did not repeat the technical review for this book, I wish to thank the original technical review team: Paul Grant, Jeff Smith, and Holly Sweeny.

I wish to thank the people who corrected my grammar mistakes and misspellings: Suzanne Morgen and Catherine Connolly. Without these people, I'd probably be getting a lot of emails from readers correcting my mistakes.

Finally, I extend my love and gratitude to my wife, Jan. She encouraged me throughout the writing process and even took the picture for the back cover.

1.

Descriptive and Inferential Statistics

Overview

Many people have a misunderstanding of what statistics entails. The trouble stems from the fact that the word "statistics" has several different meanings. One meaning relates to numbers such as batting averages and political polls. When I tell people that I am a statistician, they assume that I am good with numbers. Actually, without a computer I would be lost.

The other meaning, the topic of this book, is to **describe collections of numbers** such as test scores and to describe properties of these numbers. This subset of statistics is known as **descriptive statistics**. Another subset of statistics, **inferential statistics**, takes up a major portion of this book. One of the goals of inferential statistics is to determine whether your experimental results are "statistically significant." In other words, what is the probability that the result that you obtained could have occurred by chance, rather than an actual effect?

Descriptive Statistics

I am sure that every reader of this book is already familiar with some aspects of descriptive statistics. From early in your education, you were assigned a grade in a course, based on your average. Averages (there are several types) describe what statisticians refer to as **measures of location** or **measures of central tendency**. Most basic statistics books describe three indicators of location: the mean, median, and mode. To compute a mean, you add up all the numbers and divide by how many numbers you have. For example, if you took five tests and your scores were 80, 82, 90, 96, and 96, the mean would be (80 + 82 + 90 + 96 + 96)/5 or 88.8. To compute a median, you arrange the numbers in order from lowest to highest and then find the middle number. This number is the median. Half the numbers will be below the median and half of the numbers will be above the median. In the example of the five test scores (notice that they are already in order from lowest to highest), the median is 90. If you have an even number of numbers, one method of computing the median is to average the two numbers in the middle. The last measure of central tendency is called the mode. It is defined as the most frequent number. In this example, the mode is 96 because it occurs more than any other number. If all the numbers are different, the mode is not defined.

Besides knowing the mean or median (the mode is rarely used), you can also compute several measures of dispersion. **Dispersion** describes how spread out the numbers are. One very simple measure of dispersion is the range, defined as the difference between the highest and lowest value. In the test score example, the range is 96 – 80 = 16. This is not a very good indicator of dispersion because it is computed using only two numbers—the highest and lowest value.

The most common measure of dispersion is called the **standard deviation**. The computation is a bit complicated, but a good way to think about the standard deviation is that it is similar to the average amount each of the numbers differs from the mean, treating each of the differences as a positive number. The actual computation of a standard deviation is to take the difference of each number from the mean, square all the differences (that makes all the values positive), add up all the squared differences, divide by the number of values, minus one, and then take the square root of this value. Because this calculation is a lot of work, we will let the computer do the calculation rather than doing it by hand.

Figure 1.1 below shows part of the output from SAS when you ask it to compute descriptive statistics on the five test scores.

Basic Statistical Measures			
Location		Variability	
Mean	88.80000	Std Deviation	7.56307
Median	90.00000	Variance	57.20000
Mode	96.00000	Range	16.00000
		Interquartile Range	14.00000

Figure 1.1: Example of Output from SAS Studio

This shows three measures of location and several measures of dispersion (labeled Variability in the output). The value labeled "Std Deviation" is the standard deviation described previously, and the range is the same value that you calculated. The variance is the standard deviation squared, and it is used in many of the statistical tests that we discuss in this book.

Descriptive statistics includes many graphical techniques such as histograms and scatter plots that you will learn about in the chapter on SAS Studio descriptive statistics.

Inferential Statistics

Let's imagine an experiment where you want to test if drinking regular coffee has an effect on heart rate. You want to do this experiment because you believe caffeine might increase heart rate, but you are not sure. To start, you get some volunteers who are willing to drink regular coffee or decaf coffee and have their heart rates measured. The reason for including decaf coffee in the experiment is so that you can separate the placebo effect from a possible real effect. Because some of the volunteers might have a preconceived notion that coffee will increase their heart rate, their heart rate might increase because of a psychological reason, rather than the chemical effect of caffeine in the coffee.

You divide your 20 volunteers into two groups—to drink regular or decaf coffee. This is done in a random fashion, and neither the volunteers nor the person measuring the heart rates knows whether the person is drinking regular or decaf coffee. This type of experiment is referred to as a **double-blind, placebo-controlled, clinical trial**. We will discuss this design and several others in the next chapter.

Suppose the mean heart rate in the regular coffee group is 76 and the mean heart rate in the decaf (placebo) group is 72. Can you conclude that caffeine increases heart rate? The answer is "maybe." Why is that? Suppose that caffeine had no effect on heart rate (this is called the **null hypothesis**). If that were true, and you measured the mean heart rate in two groups of 10 subjects, you would still expect the two means to differ somewhat due to chance or natural variation. What a statistical test does is to compute the probability that you would obtain a difference as large or larger than you measured (4 points in this example) by chance alone if the null hypothesis were true. Statisticians typically call a difference statistically significant if the probability of obtaining the difference by chance is less than 5%. Be really careful here. The term significant is also used by non-statisticians to mean important. In a statistical setting, significant only means that the probability of getting the result by chance is less than 5% or some other number that you specified before the experiment began. Because statisticians like to use Greek letters for lots of things, the predefined **significance level** is called alpha (α).

Now for some terminology: You already know about the null hypothesis and a significance level. If caffeine had no effect on heart rate, what is the probability that you would see a difference of 4 points or more by chance alone? This probability is called the ***p*-value**. If the p-value is less than alpha, you reject the null hypothesis and accept the alternate hypothesis. The **alternate hypothesis** in this example is that caffeine does affect a person's heart rate. Most studies will reject the null hypothesis whether the difference is positive or negative. As strange as this sounds, this means that if the decaf group had a significantly higher heart rate than the regular coffee group, you would also reject the null hypothesis. The reason you ran this experiment was that you expected that caffeine would increase heart rate. If this was a well-established fact, supported by several clinical trials, there would be no need to conduct the study—if the effect of caffeine on heart rate was never tested, you need to account for the possibility that it could either increase or decrease heart rate. Looking for a difference in either direction is called a **2-tailed test** or a **non-directional test**.

As you saw in this example, it is possible for the null hypothesis to be true (caffeine has no effect on heart rate) and, by chance, you reject the null hypothesis and say that caffeine does affect heart rate. This is called a **type I error** (a **false positive** result). If there is something called a type I error, there is probably something called a **type II error**—and there is. This other error occurs when the alternate hypothesis is actually true (caffeine increases heart rate) but you fail to reject the null hypothesis. How could this happen? The most common reason for a type II error is that the experimenter did not have a large enough **sample** (usually a group of subjects). This makes sense: If you had only a few people in each group, it is easy to see that the means of the two groups would be different. If you had several thousand people in each group, and caffeine really had an effect on heart rate, you would be pretty sure of confirming the fact. Just as you set a significance level before you started the experiment (the 5% value that statisticians call alpha), you can also compute the probability that you make the wrong decision and claim that caffeine has no effect on heart rate when it really does (a **false negative**). The probability of a type II error is called beta (β), (more Greek).

Rather than think about the probability of a false negative, it is easier to think about the probability of a true positive. This probability is called **power**, and it is computed as 1 – beta. The last chapter of this book shows how to compute power for different statistical tests. Typically, the only way to increase power is to have a larger sample size.

Before we leave this chapter, there are two more terms that you should be familiar with. Remember in our coffee experiment, we assigned people to drink regular or decaf coffee. The 10 people in each group is called a **sample**. When you do your experiment and state your conclusion, you are not merely making a statement about your sample. You are saying that anyone who drinks regular coffee will have a higher heart rate that you estimate to be 4 points. You are making a statement about anyone who drinks regular or decaf coffee, and the name given to this theoretical group of people that you are making conclusions about is called a **population**. In practice, you define your population (everyone in the US or the world, for example), you take samples from this population, and make inferences about the effect your intervention on an outcome. That is why this branch of statistics is called **inferential statistics**.

Summary of Statistical Terms

- Measures of central tendency – statistics such as a mean or median that describe the center of a group of data values.
- Dispersion – a measure that describes how spread out the data values are. The most common measure of dispersion is the standard deviation.
- Sample – the group of subjects on whom you are conducting your experiment.
- Population – a theoretical group of subjects on whom you make inferences, based on the results from your sample.
- Type I error – a false positive result from a study. For example, concluding a drug or treatment works when it does not.
- Alpha (α) – the significance level. It is the probability that you are willing to accept for a type I error, usually set at .05.
- *p*-value – the probability of making a false positive (type I) error. If the *p*-value is less than alpha, you declare the results as significant. Some researchers prefer to just list the p-value and not set a specific alpha level.
- Type II error – a false negative result. For example, you have a drug or treatment that works but the results from the study is not significant (the probability of obtaining your result is greater than alpha.)

- Beta (β) – the probability of getting a false negative (type II) result.
- Power – the probability of a positive result when there truly is an effect. For example, you claim that your drug or treatment is better than a placebo or standard treatment, and you are correct in this decision. The power is the probability of the study rejecting the null hypothesis (finding the effect). Typically, powers of 80% or higher are considered acceptable. Large expensive studies might require powers of 90% or even 95%.

2.

Study Designs

Introduction

Whether you are reading a study or designing your own, one of the most important factors that you should consider is the study design. For example, are you going to randomly assign subjects to different treatments, or are you simply going to observe outcomes in people who have some trait in common? This chapter describes some of the commonly used study designs, and this discussion will help you decide the quality of study results and how much faith you should have in the results, regardless of the *p*-value or size of the effect.

Double-Blind, Placebo-Controlled Clinical Trials

This is the "gold standard" of all study designs. Even though the term "clinical" is in the title, this study design can be used in all fields of study. The caffeine study described in the previous chapter is an example of this study design. The first step is to select a representative sample of people from the population on which you want to make inferences. The population might be limited by age, gender, or ethnicity. For example, for your caffeine study, you would probably not include very young children. After you conduct an analysis to determine how many subjects you should include in your study (discussed in Chapter 15), you randomly assign the subjects into as many groups as necessary. This design does not necessarily have to include a placebo group even though the study design includes the word "placebo" in the title. It might not be ethical to include a placebo group if this could cause harm to the subjects in this group. For example, if you are comparing drugs to lower blood pressure in a population of subjects with high blood pressure, you might choose to have one group take one of the standard medicines and the other groups take one or more new treatments that you want to study.

Next, we need to discuss the term **double-blind**. Double-blind means that neither the subject nor the person evaluating the subject knows what treatment or drug a subject is receiving. This is quite easy to do if the subject is taking a drug in pill form. For example, the placebo group can be given an inert substance in a capsule. However, what about a surgical technique or acupuncture? How can the subject be blinded in that case? One study, done in the early 1960s, was designed to test if freezing the stomach with a balloon filled with liquid nitrogen would cure stomach ulcers. One group of subjects had their stomachs frozen. For the placebo group, the liquid nitrogen was introduced into the endoscopy tube, but then shunted out before it reached the stomach. Therefore, it looked like the real thing to the subject and to the staff (except for the person who was switching the shunt on or off). Before this study was conducted, stomach freezing was believed to help cure ulcers. However, the results of the study showed it did not work at all.

The placebo effect can be so strong that even different types of placebos can produce different results. For example, placebo injections, capsules, or pills, can produce different results. Even the color of the pill can sometimes make a difference.

Cohort Studies

A **cohort** is a group of people who share a common characteristic, such as taking vitamins or getting annual physical exams. A great deal of knowledge can be gained by comparing groups of subjects who differ in their lifestyles or other factors. For example, you could look at people who exercise regularly versus people who are more sedentary and compare endpoints such as blood pressure, cholesterol, or death from cardio-vascular disease. The advantage of studying cohorts is that you can compare groups containing thousands of subjects.

Farmington, Massachusetts is the site of one of most famous series of cohort studies ever conducted. This town was deemed to be a good representation of the US population. All types of surveys were (and still are) administered to people in the town who were willing to participate in the study. Physical exams and blood tests were also performed on a regular basis. Studies such as these gave epidemiologists data to determine how risk factors such as high cholesterol or high blood pressure are related to heart disease.

However, there are several problems with cohort studies. Because the groups are not randomly assigned to a treatment, the groups might differ on many other factors besides the one being studied. One of the most famous cohort studies that demonstrates how bias can affect the results compared women who took hormone replacement therapy (HRT) with women who did not. The results showed that women who took HRT were significantly less likely to experience a heart attack or stroke.

Years later, a large, double-blind, placebo-controlled clinical trial was conducted to compare HRT to a placebo. The results showed that HRT increased the likelihood of heart attacks and strokes. Why were the results of the earlier studies and the new study so different? Who were the women in the early study who received HRT? Most likely, they were more affluent, had better medical care, and might have lived a healthier lifestyle. It was because of those reasons that it looked like HRT was beneficial when it was not. This demonstrates that while cohort studies can be very useful, the results need to be looked at with more skepticism than results from clinical trials.

Case-Control Studies

The last study design to be described in this chapter is a case-control study. **Cases** are subjects who already have a disease or condition of interest. **Controls** are subjects chosen to be similar to the cases except for the fact that they do not have the disease or condition in question. Typical case-control studies compare the two groups to see whether there are differences in lifestyle, exposure to chemicals (such as cigarette smoke), or medications. One such study compared women who gave birth to babies with a condition called neural tube defects. This condition causes an abnormal brain, spine, or spinal column. Controls were women who gave birth to babies without this condition. The two groups were compared on many factors. One factor that showed a significant difference was between women who took folic acid supplements and those who did not.

Why not do a clinical trial to test if folic acid is useful in preventing neural tube defects? There are several reasons why a clinical trial would be difficult. First, neural tube defects are quite rare, so you would need to recruit thousands of women for the study. Another problem is an ethical one. If doctors believe that folic acid helps prevent this condition, how can they, in good conscience, deny folic acid to a placebo group?

Case-control studies are very useful when you are studying a rare condition. However, the results from a case-control study have to be looked at with caution. In the neural tube defect study, women who gave birth to babies with this condition might have had more vivid recollections of what they ate and did while pregnant, compared to women who had normal babies. This problem is called **recall bias**, and it is difficult to control for. Finally, because you start with people with the disease in one group and healthy people in the other group, you cannot compute incidence rates (how many people get the disease or develop the condition in a certain period of time).

Conclusion

The three study designs described in this chapter are all useful in comparing drugs or treatments in groups of people. There are times when you cannot blind the treatment. One example of this is if you are comparing two different methods to teach students to read. Some of the children (depending on age) might realize they are being taught differently than their friends. The results of studies that are not blinded can still provide useful information.

3.

What Is SAS OnDemand for Academics?

An Overview of SAS OnDemand for Academics (SAS in the Cloud)

SAS is many things: A data analysis tool, a programming language, a statistical package, a tool for business intelligence, and more. Until recently, you could get access to SAS by paying a license fee (this could be an individual license, or a license purchased by a company for as many users as necessary).

The really big news is that anyone can now obtain SAS for FREE! It's called SAS OnDemand for Academics (ODA)—but don't be fooled by the name. **Anyone can use this free version of SAS, not just students enrolled in a class**.

Some of you might know about a SAS product called the SAS University Edition. This was another free version of SAS, but you had to download software to create a virtual computer on your real computer, then download the SAS software, and finally, set up a way to read and write files from your "real" computer to the "virtual computer." This caused many people massive headaches (including this author). The great news about SAS ODA is that **you don't have to download anything!** You access SAS on a cloud platform. Also, reading data from your real computer is quite simple.

And now for the caveats: This product was developed so that people can use it to learn how to program and run tasks using SAS. It is not supposed to be used for commercial purposes. One final note: there is a 5-gigabyte limit for data files. That is certainly not a problem for learning how to use SAS.

On many college campuses, students taking statistics courses or any course that needs a powerful analytic tool could access a computer language called R for free. Since free is better than not free, these institutions sometimes choose to use R instead of SAS. That is fine, except that when these students graduate, they find that in the corporate world, SAS is by far the major package for powerful statistical analysis, data manipulation, and reporting. By offering a free version of SAS, users now have a choice between SAS or R, and SAS Institute is hoping that the majority of users choose SAS.

ODA uses SAS Studio as the interface. SAS Studio provides an environment that includes a point-and-click facility for performing many common tasks, such as producing reports, graphs, data summaries, and statistical tests. For those who either enjoy programming or have more complicated tasks, SAS Studio also enables you to write and run your own programs.

Registering for ODA

To gain access to ODA, you need to register. Part of the registration process is to create a SAS profile. If you already have a SAS profile, skip that portion of the instructions.

To start, point your browser to:

https://welcome.oda.sas.com

and follow the clear directions. If you do not already have a SAS Profile, you will be asked to create one. It is free. Once you have a profile, you can enter the URL above and see the following screen.

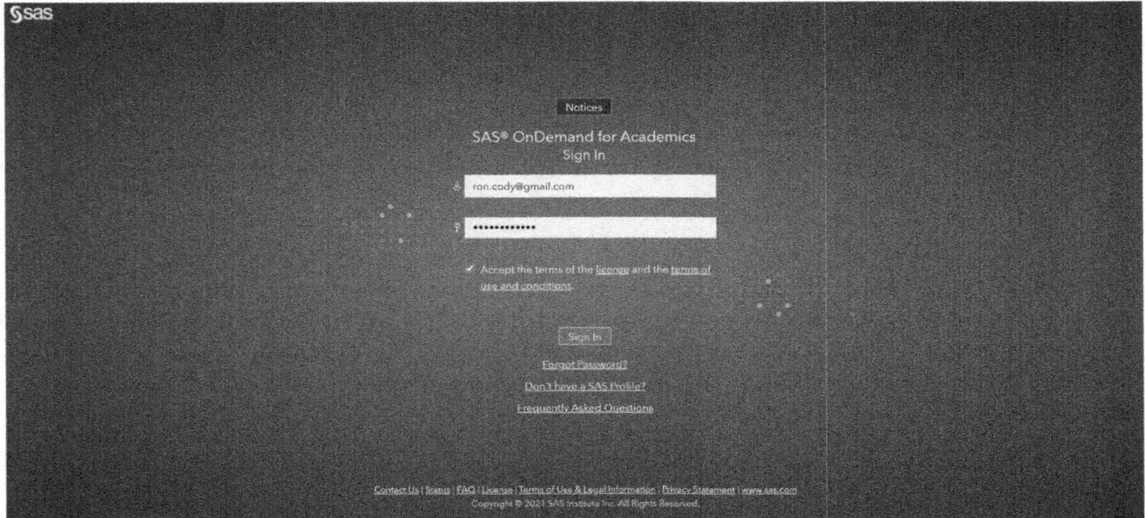

Figure 3.1: Opening Screen for SAS OnDemand for Academics

Be sure to check the box "Accept the terms" and then click Sign In.

Once you sign in, you will see the following.

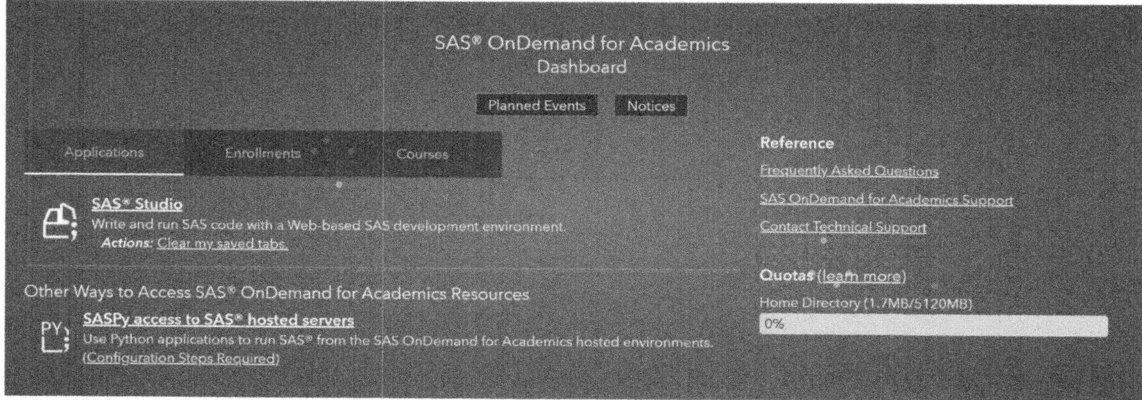

Figure 3.2: Click SAS Studio to Get Started

Click SAS Studio and you are ready to run tasks, write programs, or both. Here is the SAS Studio opening screen.

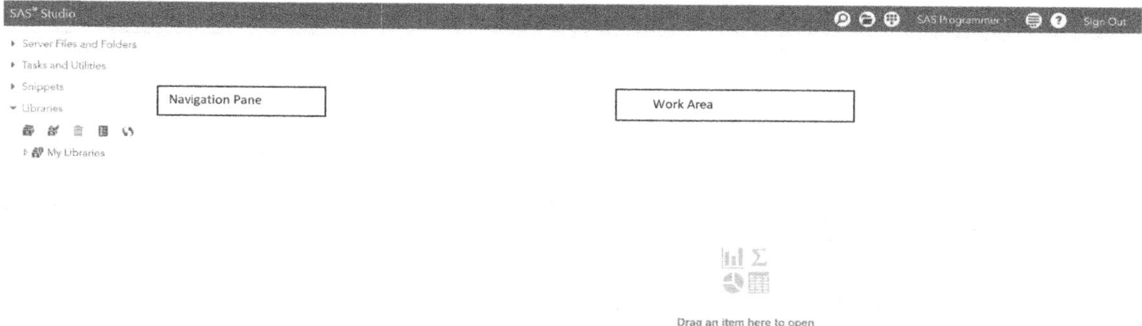

Figure 3.3: Opening SAS Studio Screen

The left side of the screen is called the navigation panel–the right side of the screen is the work area (where you can write your own programs). Your screen will not look exactly like Figure 3.3 because this author has already created several folders.

If you were planning to write a program you would click the icon circled in Figure 3.4 (shown below).

Figure 3.4: Preparing to Write a Program

Your screen would then look like this:

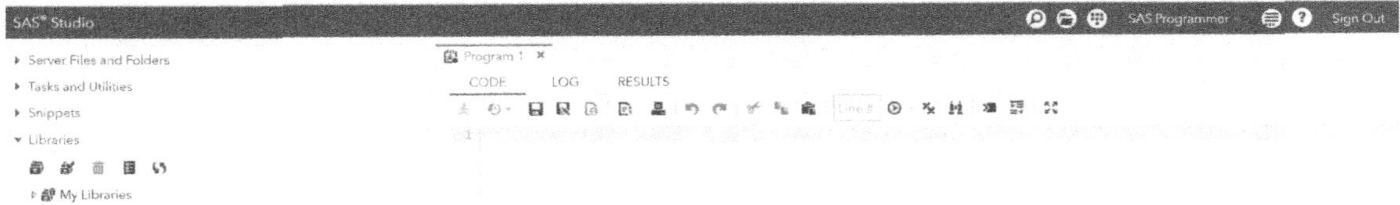

Figure 3.5: Preparing to Write a New Program

The right side of the screen now shows the three tabs: CODE, LOG, and RESULTS. The Work area is where you write your own programs. The next chapter will show you how to use some of the built-in tasks. Once you are familiar with how SAS Studio tasks work, you will move on to uploading your own data to the SAS Studio cloud and performing some basic tasks.

Conclusion

SAS OnDemand for Academics is the perfect tool for analyzing data using built-in tasks and/or learning how to program in SAS. For those readers who previously used SAS University Edition (which is being phased out), the transition to ODA is smooth and simple.

The next few chapters discuss some of the built-in data sets that you can use to perfect your skills, as well as instructions for using your own data, either on Excel spreadsheets (a very common data source) or comma-separated values (CSV) files.

4.

SAS Studio Tasks

Introduction

Now that you have gained access to SAS Studio in the cloud (my preferred term for ODA), you can start exploring some of the built-in tasks. SAS includes a few hundred data sets in a folder called SASHELP that you can use to get familiar with SAS Studio tasks or you can even use the sample data to write and test your own programs. This chapter will explore how some of the tasks work. Once you have run a few tasks, you should have the skills to run any of the other SAS Studio tasks.

Figure 4.1 shows an enlarged view of the Navigation pane. Notice the tabs such as Server Files and Folders or Libraries. You see a small triangle to the left of each task. If you click any tab, it expands to show subtasks. The order of the tasks also change position (usually come to the top) when it is selected.

Exploring the SAS Studio Libraries

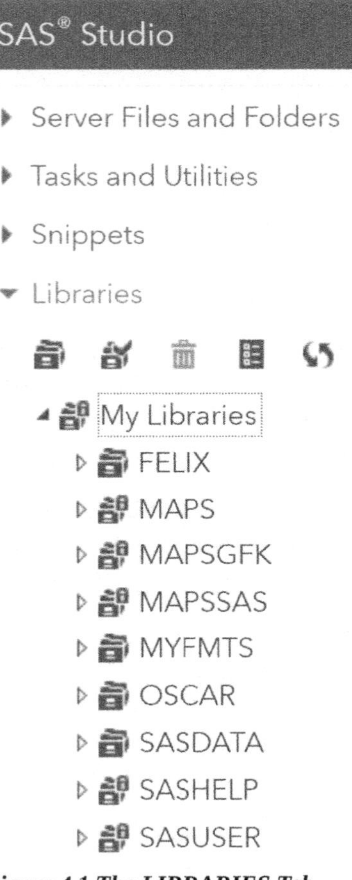

Figure 4.1 The LIBRARIES Tab

In Figure 4.2, the SASHELP tab was selected. Notice the triangle now points downward. When you click a particular library (SASHELP in this example), the triangle now points down and to the right and the list of all the SASHELP libraries appears.

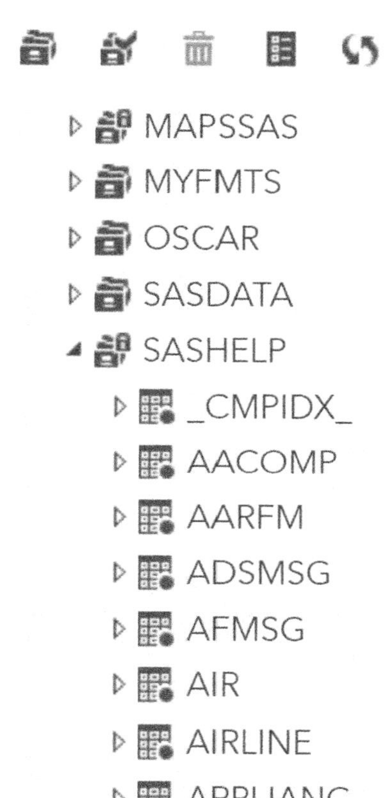

Figure 4.2: Selecting the SASHELP Libraries

Figure 4.3 shows data sets farther down the list. The Retail data set was selected. You can either double click the data set or drag it to the work area.

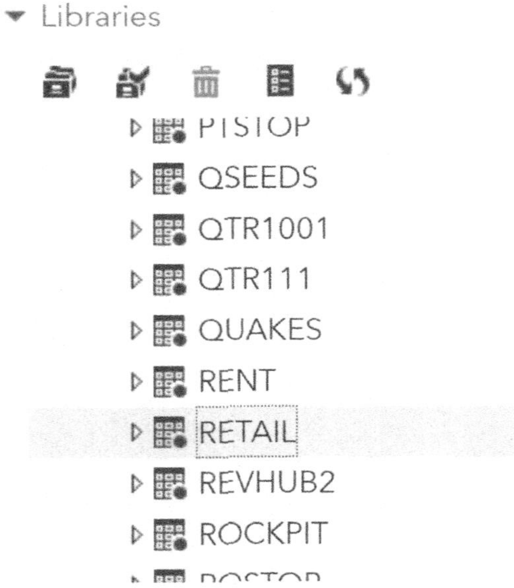

Figure 4.3: Selecting the Retail Data Set

On the left of Figure 4.4, you see a list of variables in the data set–on the right, a spreadsheet-like view of the individual data values. There are vertical and horizontal scroll bars that you can use to move around the rows and columns.

SASHELP.RETAIL ×

View: Column names Filter: (none)

Columns

- Select all
- SALES
- DATE
- YEAR
- MONTH
- DAY

Variables

Total rows: 58 Total columns: 5 Rows 1-58

	SALES	DATE	YEAR	MONTH	DAY
1	$220	80Q1	1980	1	1
2	$257	80Q2	1980	4	1
3	$258	80Q3	1980	7	1
4	$295	80Q4	1980	10	1
5	$247	81Q1	1981	1	1
6	$292	81Q2	1981	4	1
7	$286	81Q3	1981	7	1
8	$323	81Q4	1981	10	1
9	$284	82Q1	1982	1	1
10	$307	82Q2	1982	4	1
11	$318	82Q3	1982	7	1
12	$343	82Q4	1982	10	1

Spreadsheet Like View

Figure 4.4: "Worksheet" View of the Retail Data Set

Manipulation How Data Values Are Displayed

If you right-click a column head, a list of actions appears (Figure 4.5). For this example, you have decided to sort the data in ascending (low to high) order. One of the other very useful options is to Size grid columns to content. This action makes the table more compact.

Total rows: 58 Total columns: 5

	SALES ▲	DATE	YEAR	MONTH	DAY
1	$220	80Q1			
2	$247	81Q1			
3	$257	80Q2			
4	$258	80Q3			
5	$284	82Q1			
6	$286	81Q3	1981	7	1
7	$292	81Q2	1981	4	1

- Sort Ascending
- Sort Descending
- Sort by Data Order
- Add Filter
- Size grid columns to content
- Restore original column widths

Figure 4.5: Actions You Can Perform on a Column

Figure 4.6 show the table sorted by the variable Year. (Note: this is a partial listing.)

Total rows: 58 Total columns: 5

	SALES	DATE	YEAR ▲	MONTH	DAY
1	$220	80Q1	1980	1	1
2	$295	80Q4	1980	10	1
3	$258	80Q3	1980	7	1
4	$257	80Q2	1980	4	1
5	$292	81Q2	1981	4	1
6	$286	81Q3	1981	7	1
7	$247	81Q1	1981	1	1
8	$323	81Q4	1981	10	1
9	$343	82Q4	1982	10	1
10	$318	82Q3	1982	7	1

Figure 4.6: Data Sorted by Year

Exploring SAS Studio Tasks

There are other actions that you can perform in a similar manner. However, using the built-in tasks usually provides a neater view of the data. Figure 4.7 is a list of the built-in SAS Studio tasks.

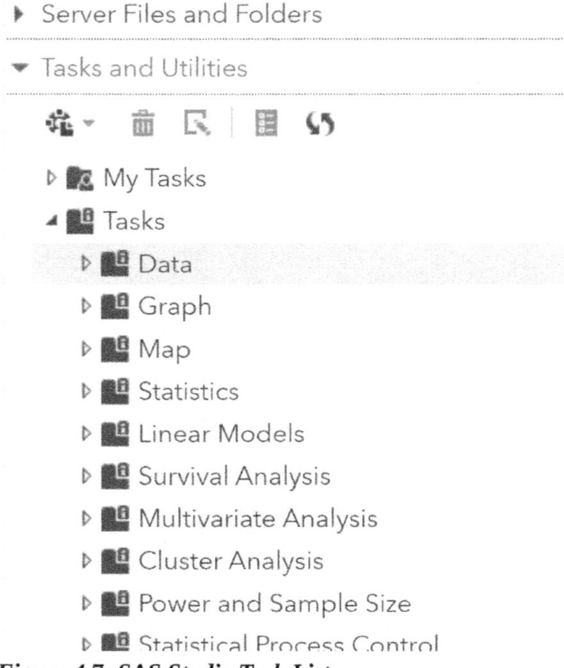

Figure 4.7: SAS Studio Task List

Figure 4.8 resulted from this author selecting the Data task. Notice the triangle to the left of Data now points downward, and you can see all of the tasks grouped under Data. (What you see in the figure is a partial list of Data tasks.)

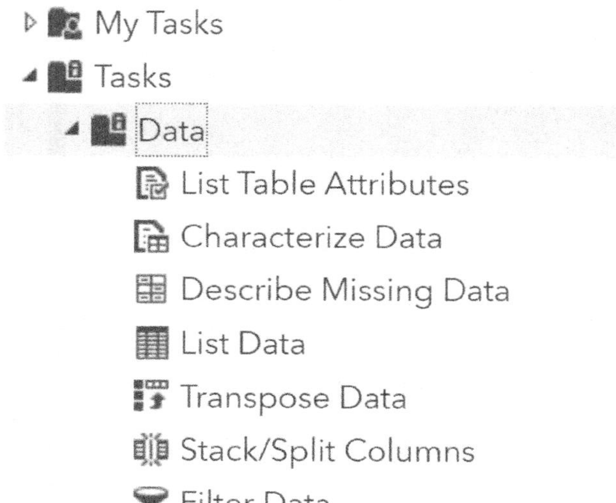

Figure 4.8: Partial List of Data Tasks

Let's use Characterize Data in the Data task list as a demonstration. This task requires similar information to many of the other tasks, so it makes for a good example. Double-click Characterize Data to begin.

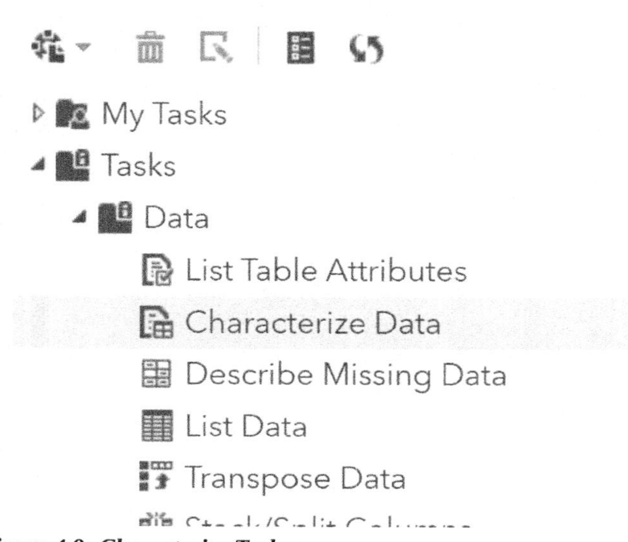

Figure 4.9: Characterize Task

Many of the tasks have a series of tabs. For this task, you see DATA, OPTIONS, and INFORMATION tabs. You typically start out with the DATA tab. It is here where you specify which SAS data set you want to process. The icon circled in Figure 4.10 is the one that lets you choose a library and a data set, as well as a list of variables that you want to explore.

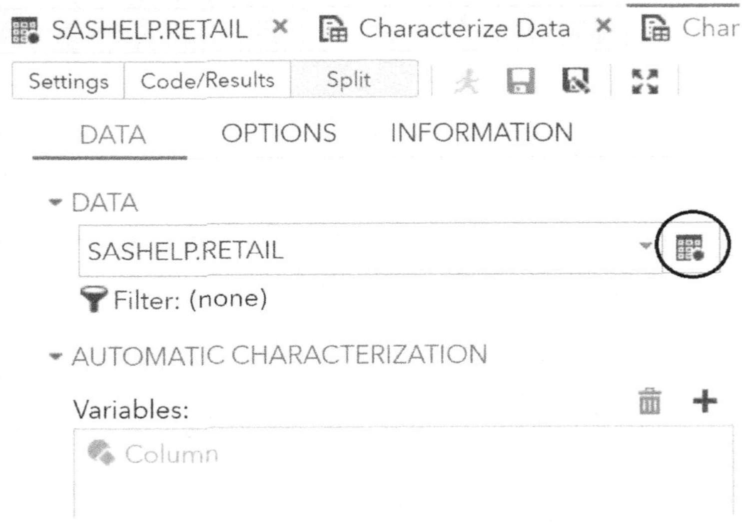

Figure 4.10: Selecting a Data Set

In this example, you chose the Retail data set in the SASHELP library. If you click the plus sign (+) at the top right of the box labeled **Variables**, you will see a list of variables in the Retail data set. There are several ways to select variables (which will be familiar to people who use Windows operating systems). First, you can hold down the control key and click each variable that you want to select. If you click a variable that is already selected (shaded in gray), it will be deselected. A second way to select variables is to click a variable, hold down the Shift key, then click another variable. All variables from the first to the last will be selected. In Figure 4.11, the variables Sales, Year, Month, and Day have been selected. Once you select your variables, be sure to click the OK button at the bottom of the screen.

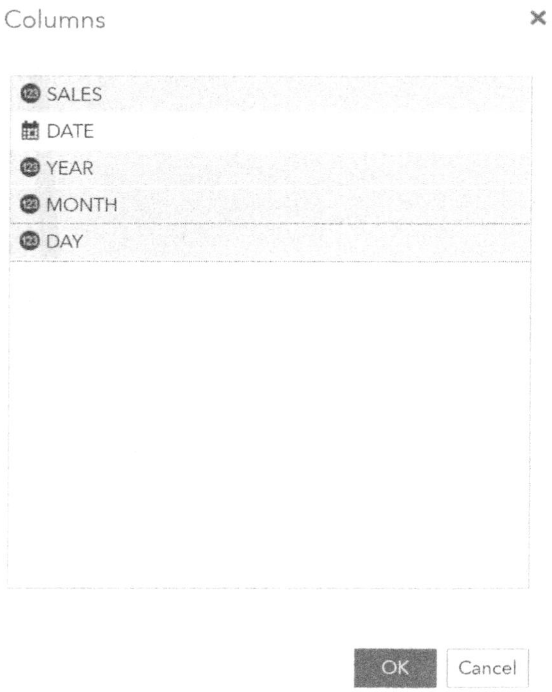

Columns ×

SALES
DATE
YEAR
MONTH
DAY

OK Cancel

Figure 4.11: Selecting Variables

Figure 4.12 shows the OPTION tab for this task. Some boxes are selected (checked) by default. You can click any selected option to deselect it. You have decisions for Categorical variables and ones for Numeric variables. Once you have selected the options you want, click the Run icon (also called the Submit icon). It is circled in Figure 4.12.

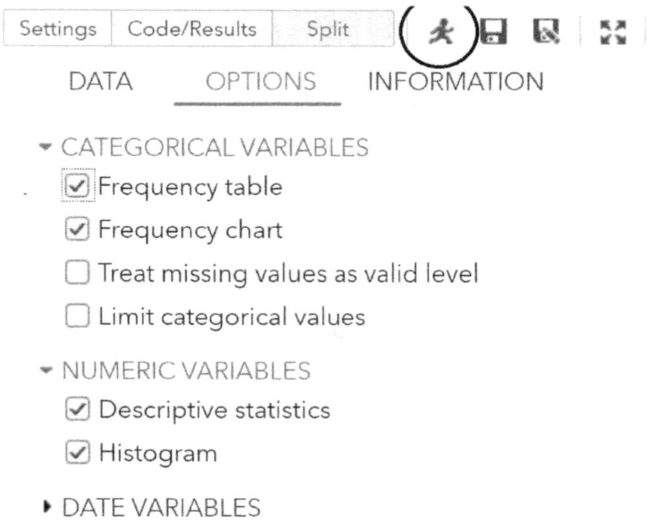

Settings | Code/Results | Split

DATA OPTIONS INFORMATION

CATEGORICAL VARIABLES

- [x] Frequency table
- [x] Frequency chart
- [ ] Treat missing values as valid level
- [ ] Limit categorical values

NUMERIC VARIABLES

- [x] Descriptive statistics
- [x] Histogram

DATE VARIABLES

Figure 4.12: Demonstrating the OPTIONS Tab

The result is shown on the right side of the screen as shown in Figure 4.13.

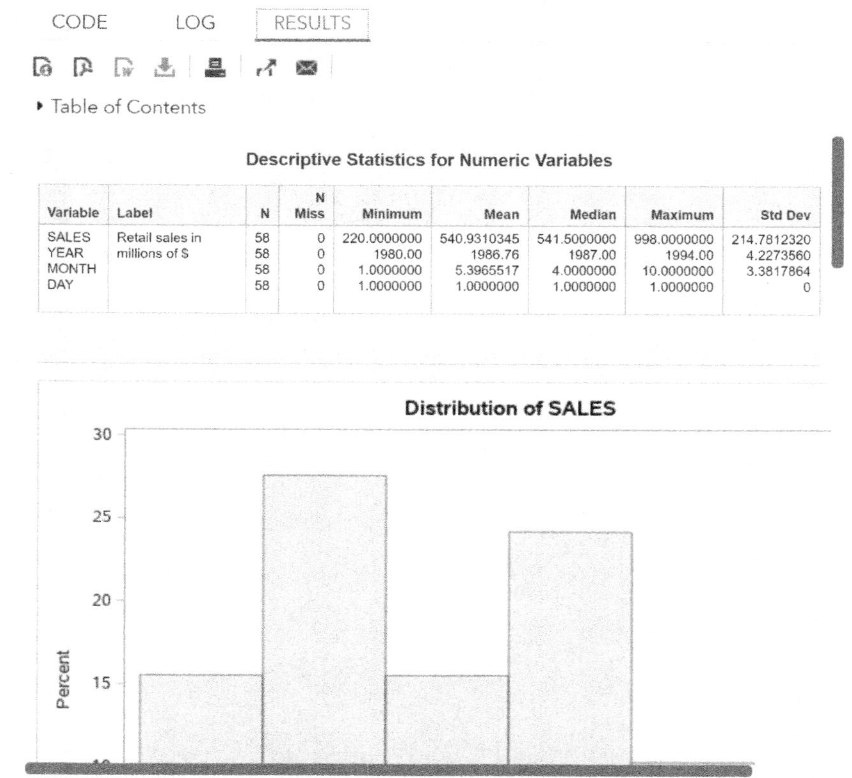

Variable	Label	N	N Miss	Minimum	Mean	Median	Maximum	Std Dev
SALES	Retail sales in millions of $	58	0	220.0000000	540.9310345	541.5000000	998.0000000	214.7812320
YEAR		58	0	1980.00	1986.76	1987.00	1994.00	4.2273560
MONTH		58	0	1.0000000	5.3965517	4.0000000	10.0000000	3.3817864
DAY		58	0	1.0000000	1.0000000	1.0000000	1.0000000	0

Figure 4.13: The Results Window

You can use the scroll bars to see the entire output. However, if you click the icon second in from the right at the top of the listing (it looks like a diagonal line with arrows at both ends)), the results are displayed in a separate browser window, as shown in Figure 4.14.

Descriptive Statistics for Numeric Variables

Variable	Label	N	N Miss	Minimum	Mean	Median	Maximum	Std Dev
SALES	Retail sales in millions of $	58	0	220.0000000	540.9310345	541.5000000	998.0000000	214.7812320
YEAR		58	0	1980.00	1986.76	1987.00	1994.00	4.2273560
MONTH		58	0	1.0000000	5.3965517	4.0000000	10.0000000	3.3817864
DAY		58	0	1.0000000	1.0000000	1.0000000	1.0000000	0

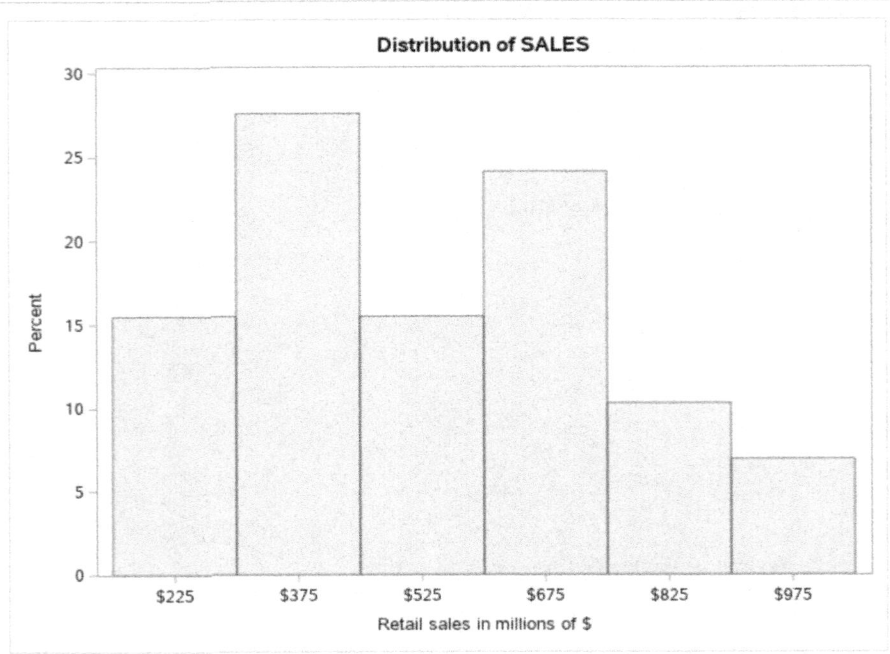

Figure 4.14: Displaying the Results in a Separate Browser Window

This makes it much easier to read.

Filtering (Subsetting) Data

Filtering (also known as subsetting) data can be useful for several reasons. Having written *Cody's Data Cleaning Techniques, 3rd Edition*, I believe that exploring your data both in tabular and graphical form should be the preliminary step before any type of analysis is attempted. You can use a filter to look for abnormally high or low data values that might be data errors. If they are not data errors, they might still affect the type of statistical analysis that you plan to perform (perhaps a nonparametric rather than a parametric test).

You will find the Filter task in the list of Data tasks. (Note, you can also filter data directly in many of the SAS Studio tasks as well). Start off by double-clicking Filter Data in the list (Figure 4.15).

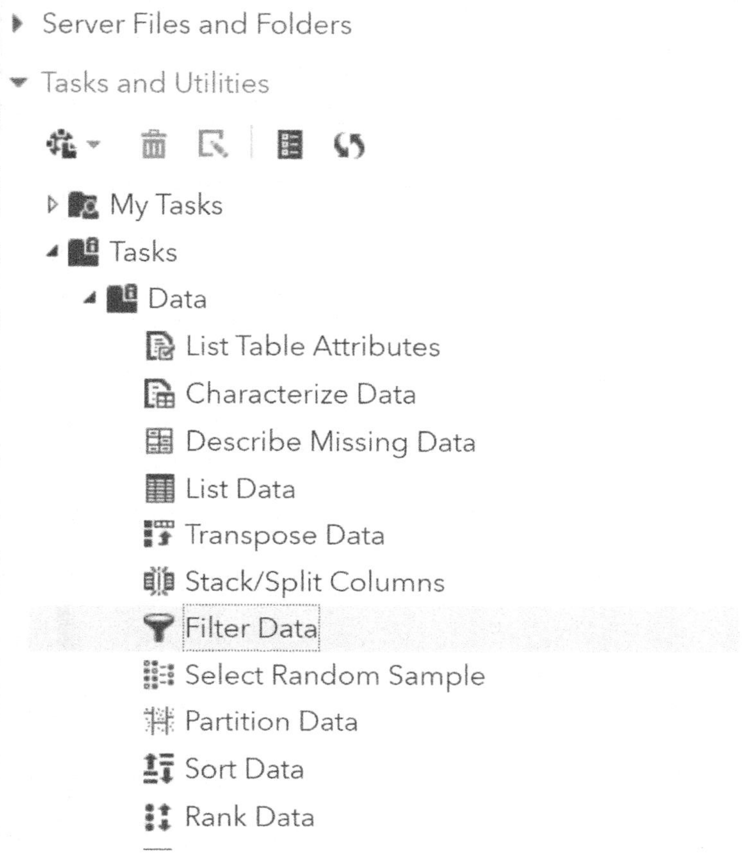

Server Files and Folders
Tasks and Utilities

My Tasks
Tasks
Data
List Table Attributes
Characterize Data
Describe Missing Data
List Data
Transpose Data
Stack/Split Columns
Filter Data
Select Random Sample
Partition Data
Sort Data
Rank Data

Figure 4.15: Choosing the Filter Data Task

For this example we are going to use the Retail data set in the SASHELP Library. The DATA tab is displayed in Figure 4.16.

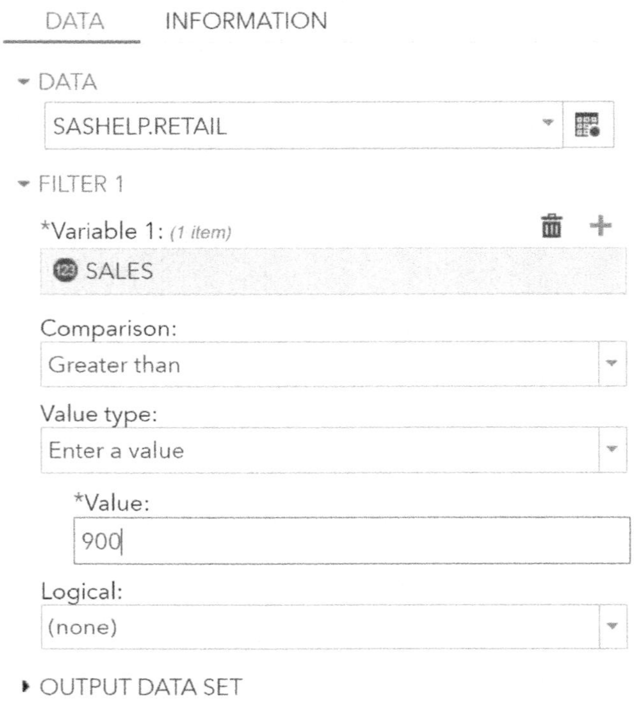

DATA INFORMATION

DATA

SASHELP.RETAIL

FILTER 1

*Variable 1: *(1 item)*

SALES

Comparison:

Greater than

Value type:

Enter a value

*Value:

900

Logical:

(none)

OUTPUT DATA SET

Figure 4.16: Displaying the DATA Tab

For this example, we are going to use the Retail data set from the SASHELP library that we have used before. Using the plus (+) sign, select a variable. In Figure 4.16, the variable Sales was chosen. Under Comparison box, there is a pull-down menu of choices:

- Less than
- Less than or equal
- Equal
- Not equal
- Greater than
- Greater than or equal

For this example, you want to see all sales greater than $900. Note that you do not include dollar signs or commas in these values. Under Value type you also have several choices. Here you can enter a value (900 in this example). If you have a character variable, the task will show you a list of distinct values from which you can choose. At this point, you can run the task and see the results, or you can create an output data set (by choosing the OUTPUT DATA SET tab). Let's run the procedure.

Total rows: 4 Total columns: 5 Rows 1-4

	SALES	DATE	YEAR
1	$909	93Q2	1993
2	$920	93Q3	1993
3	$991	93Q4	1993
4	$998	94Q2	1994

Figure 4.17: Sales Greater than $900

This author has no idea what items were for sale in the Retail data set, but there were four items over $900.

Listing Your Data

You will find that the List Data task is a task that you will use quite often. It is very flexible. You can list an entire data set, select which variable you want to include, or, my favorite, list the first "n" observations in the data set. This option is useful because you can see any mistakes in your programming or logic if, for example, you see an entire column of missing values.

The following example uses the List Data task to list the first eight observations and all of the variables in the Retail data.

Start out with the List Data task in the Data list. Choose the data set in the usual way and select all the variables. There is a box labeled "Group analysis by" but we will ignore it for now. Your screen should like Figure 4.18.

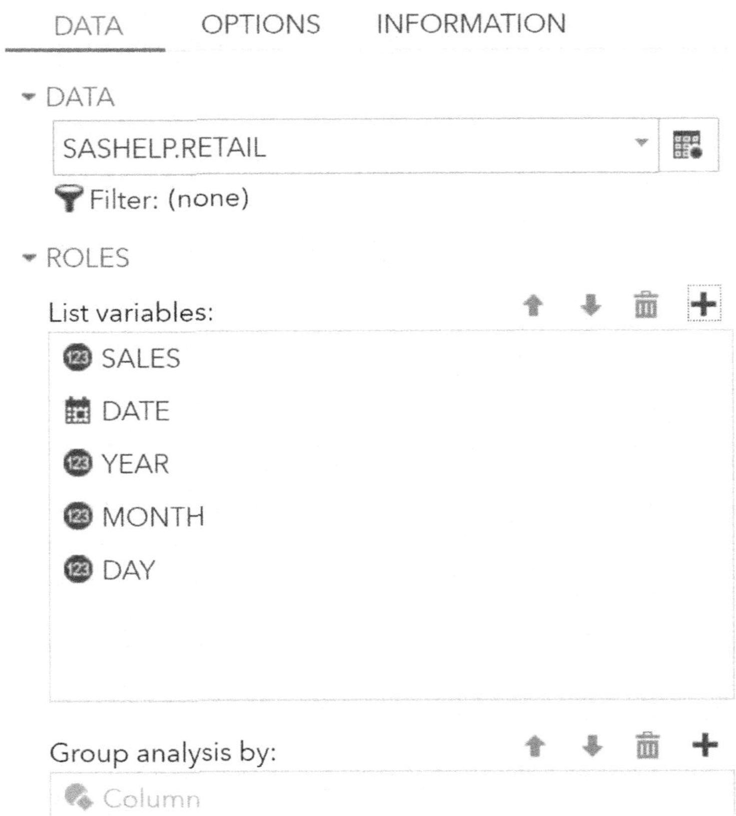

DATA OPTIONS INFORMATION

DATA

SASHELP.RETAIL

Filter: (none)

ROLES

List variables:

SALES

DATE

YEAR

MONTH

DAY

Group analysis by:

Column

Figure 4.18: Choosing Variables in the Retail Data Set

Next, click the OPTIONS tab. It looks like this:

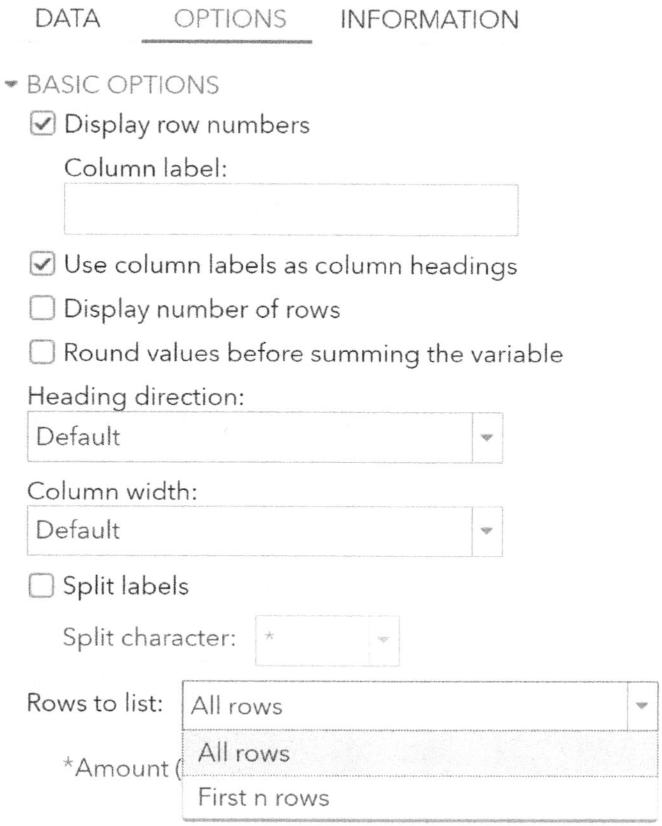

DATA OPTIONS INFORMATION

▾ BASIC OPTIONS

☑ Display row numbers

Column label:

☑ Use column labels as column headings

☐ Display number of rows

☐ Round values before summing the variable

Heading direction:

Default

Column width:

Default

☐ Split labels

Split character: *

Rows to list: All rows

*Amount (

All rows

First n rows

Figure 4.19: The Options Tab

There are quite a few OPTIONS to choose from. The box labeled Use column labels as column headings is important. Let me explain: In the "early" days of computing (and early versions of SAS), variable names were limited to eight characters, sometimes making it difficult to know what a variable name meant. SAS created labels so that you could provide a description of any or all of your variables. When you write SAS programs, you need to use variable names–when you want to give someone a report, you might prefer variable labels as column headings. By the way, in current versions of SAS, variable names can be up to 32 characters in length, so labels are less important. Notice the box at the bottom of the OPTIONS tab labeled Rows to list. The default is to list all the rows. If you choose the option First n rows, you can choose how many rows to list (see Figure 4.20).

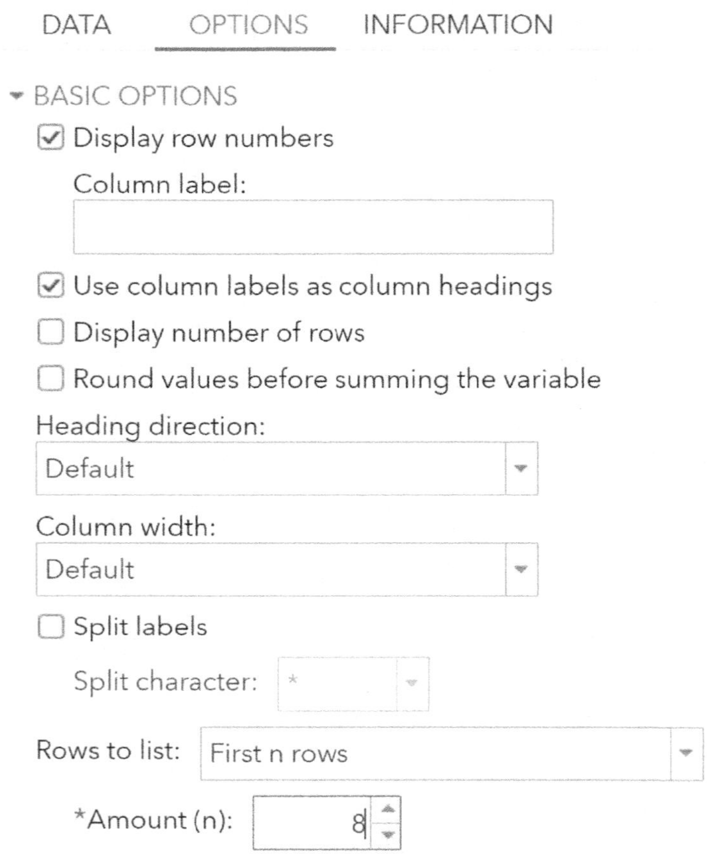

DATA OPTIONS INFORMATION

BASIC OPTIONS

☑ Display row numbers

Column label:

☑ Use column labels as column headings

☐ Display number of rows

☐ Round values before summing the variable

Heading direction:

Default

Column width:

Default

☐ Split labels

Split character: *

Rows to list: First n rows

*Amount (n): 8

Figure 4.20: OPTIONS for the List Data Task

Here is the result.

List Data for SASHELP.RETAIL

Obs	Retail sales in millions of $	DATE	YEAR	MONTH	DAY
1	$220	80Q1	1980	1	1
2	$257	80Q2	1980	4	1
3	$258	80Q3	1980	7	1
4	$295	80Q4	1980	10	1
5	$247	81Q1	1981	1	1
6	$292	81Q2	1981	4	1
7	$286	81Q3	1981	7	1
8	$323	81Q4	1981	10	1

Figure 4.21: Listing the First Eight Rows

Conclusion

In this chapter, you saw how to find built-in SAS Studio tasks, and you learned how to select data sets from the SASHELP library. The next step is to perform these operations on your own data. Chapter 5 shows you how to import data from Excel workbooks or CSV files and create SAS data sets.

5.

Importing Data into SAS

Introduction

Now that you have learned how to perform operations on built-in SASHELP data sets, it's time to see how to import your own data into a SAS library.

SAS data sets contain two parts: the first part is called the **data descriptor**, also known as **metadata**. Metadata is a fancy word for data about your data. In the case of a SAS data set, this portion of the data set contains such information as the number of rows and columns in the table, the column names, the data type for each column (SAS has only two data types—character and numeric), and other information about when the data set was created.

The second part of a SAS data set contains the actual **data values**. If you tried to examine a SAS data set using another program such as Word or Notebook, it would show up as nonsense. Only SAS can read, write, and analyze data in a SAS data set. If you have data in Excel workbooks or text files, you need to convert that data into a SAS data set before you can use SAS to modify or analyze the data.

In this chapter, you will see how easy it is to import your own data from Excel workbooks, CSV files, and many other file formats such as Microsoft Access and SPSS, and create SAS data sets.

Uploading Data from Your Local Computer to SAS Studio

Before you start uploading data files and programs to the cloud, you need to create a folder to hold the files. This is similar to folders on a Windows platform. If you open up the Server Files and Folders tab in the navigation pane, you will see the following.

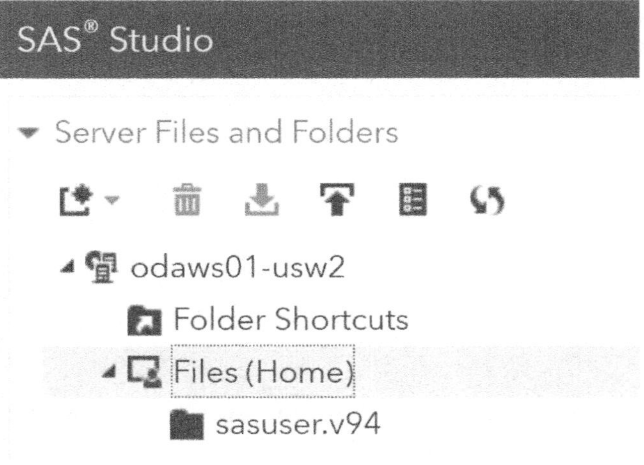

Figure 5.1: Server Files and Folders Tab

Next, click the New icon as shown circled in Figure 5.2.

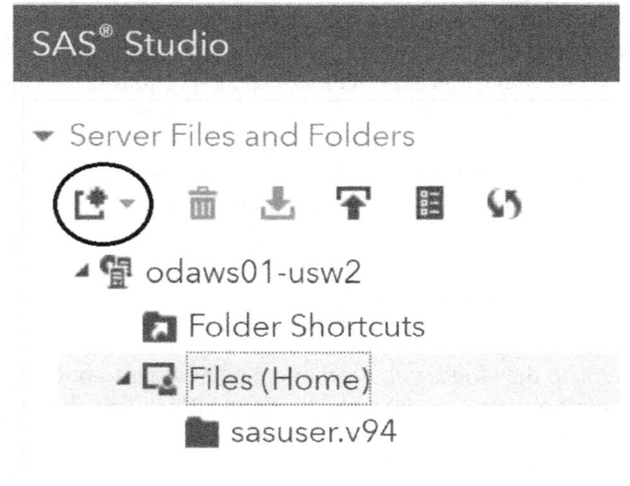

Figure 5.2: Click the New Icon

From the pull-down list, select Folder (Figure 5.3).

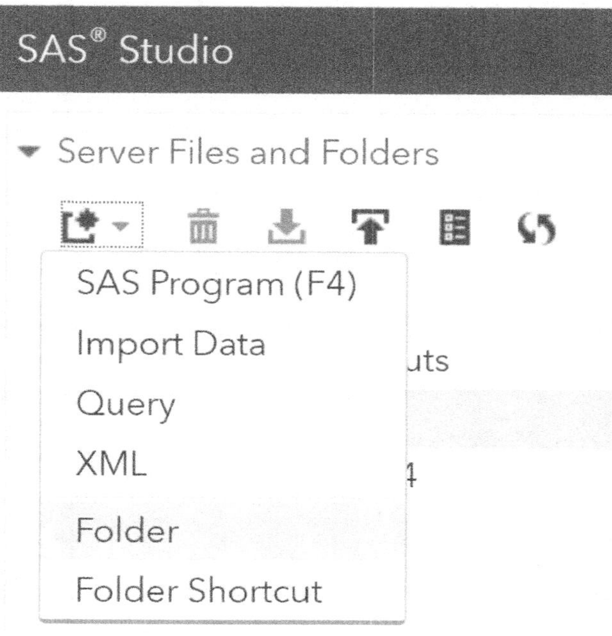

Figure 5.3: Select Folder in the Pull-Down List

You can now name your new folder. In Figure 5.4, the new folder is called Statistics Data.

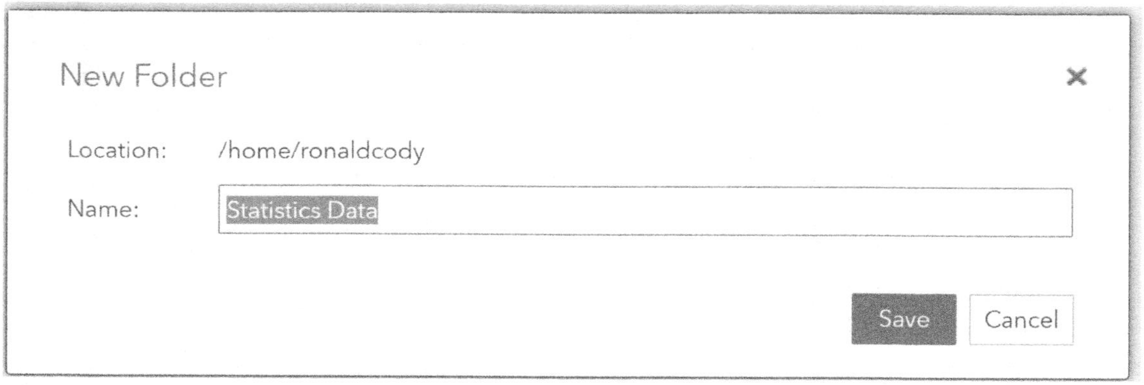

New Folder

Location: /home/ronaldcody

Name: Statistics Data

Save Cancel

Figure 5.4: Naming Your Folder

Click Save to finish creating your folder. Your folder now shows up in the list of folders as shown in Figure 5.5.

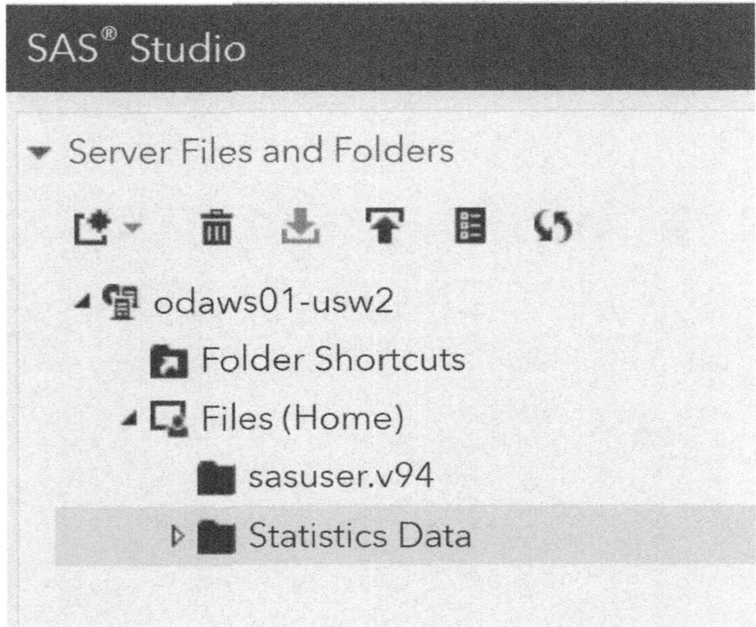

Figure 5.5: Your New Folder is Now Included in the List of Files

Notice that there is a small triangle to the left of the folder name. You can expand a list by clicking this triangle. When you expand the contents of a folder, the triangle points downward and you see the list of files. This action is a "toggle," that is, it will either expand or collapse the lists.

For this example, we are going to upload an Excel workbook called Grades.xlsx to the Statistics Data location. In Figure 5.6, you see a copy of the worksheet that contains student names, ID numbers, some Quiz and exam grades. The first row of the worksheet contains variable names (also known as column names). The remaining rows contain data on three students (yes, it was a very small class). The worksheet name was not changed, so it has the default name Sheet1.

	A	B	C	D	E	F	G	H
1	Name	ID	Quiz1	Quiz2	Midterm	Quiz3	Quiz4	Final
2	Jones	12345	88	80	76	88	90	82
3	Hildebrand	22222	95	92	91	94	90	96
4	O'Brien	33333	76	78	79	81	83	80
5								

Figure 5.6: Contents of the Spreadsheet Grades.xlsx

In order to permit SAS Studio to access data files stored on your hard drive, you first need to upload the files to SAS Studio. As shown in Figure 5.7, you click the Upload icon.

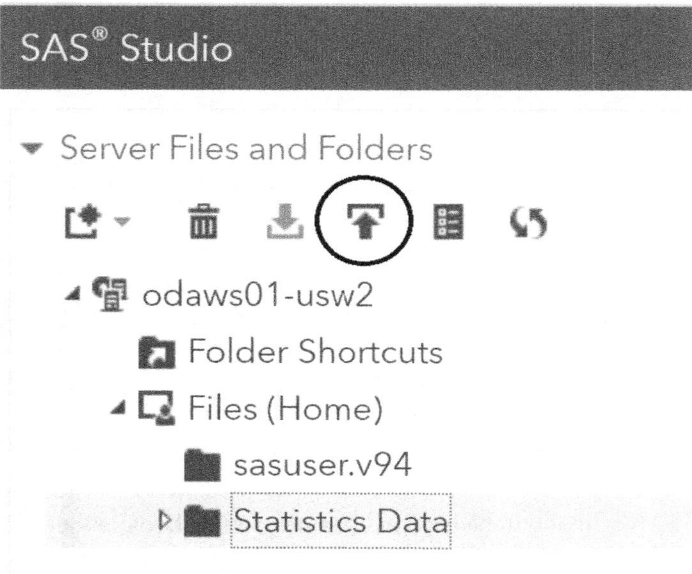

Figure 5.7: Click the Upload Icon to Upload One or More Files

This brings up a screen where you can choose which files you want to upload to SAS Studio.

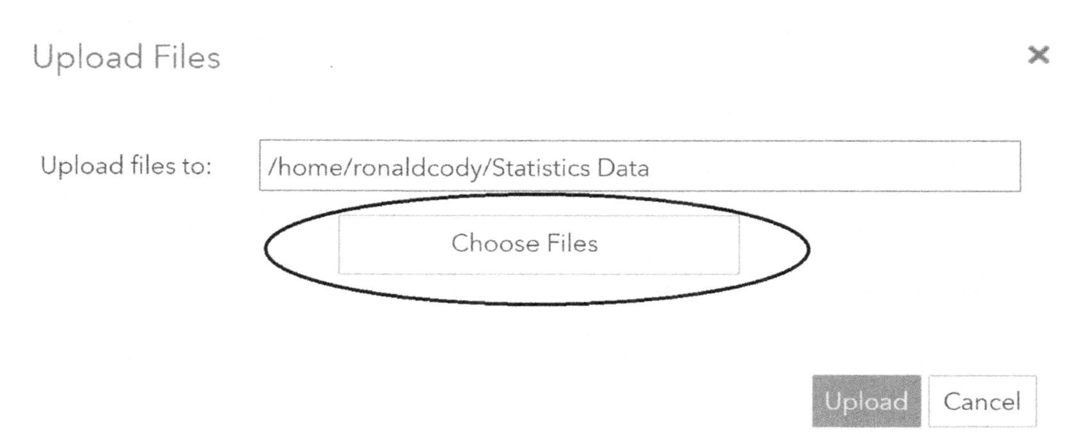

Figure 5.8: Choose Files Anywhere on Your Hard Drive

Click the Choose Files box and then you can choose a single file or multiple files. There are two ways to choose multiple files, using methods familiar to Windows users. One way is to hold down the control key and click each file that you want to upload. The other is to click one file, hold down the

Shift key, and then click another file. All files from the first to the last will be selected. In Figure 5.9 this latter method was used.

Name	Date modified	Type	Size
Count.xlsx	2/27/2021 9:23 AM	Microsoft Excel W...	8 KB
Create_Datasets.sas	2/27/2021 9:23 AM	SAS System Progr...	9 KB
Grades.xlsx	2/27/2021 9:23 AM	Microsoft Excel W...	9 KB
Perch.xlsx	2/27/2021 9:23 AM	Microsoft Excel W...	10 KB
SAT_Scores.xlsx	2/27/2021 9:23 AM	Microsoft Excel W...	9 KB
Ttest_Data.xlsx	2/27/2021 9:23 AM	Microsoft Excel W...	9 KB
Yoga.xlsx	2/27/2021 9:23 AM	Microsoft Excel W...	9 KB

Figure 5.9: Choose Your File(s)

For this example, we are uploading all the files that will be used in examples in this book. Notice the Excel workbook Grades.xlsx is included in this group of files. Once you have selected the files that you want to upload, click the box labeled Upload.

Although the maximum file space on the SAS Studio cloud is 5 megabytes, you cannot upload more than 1 megabyte at a time.

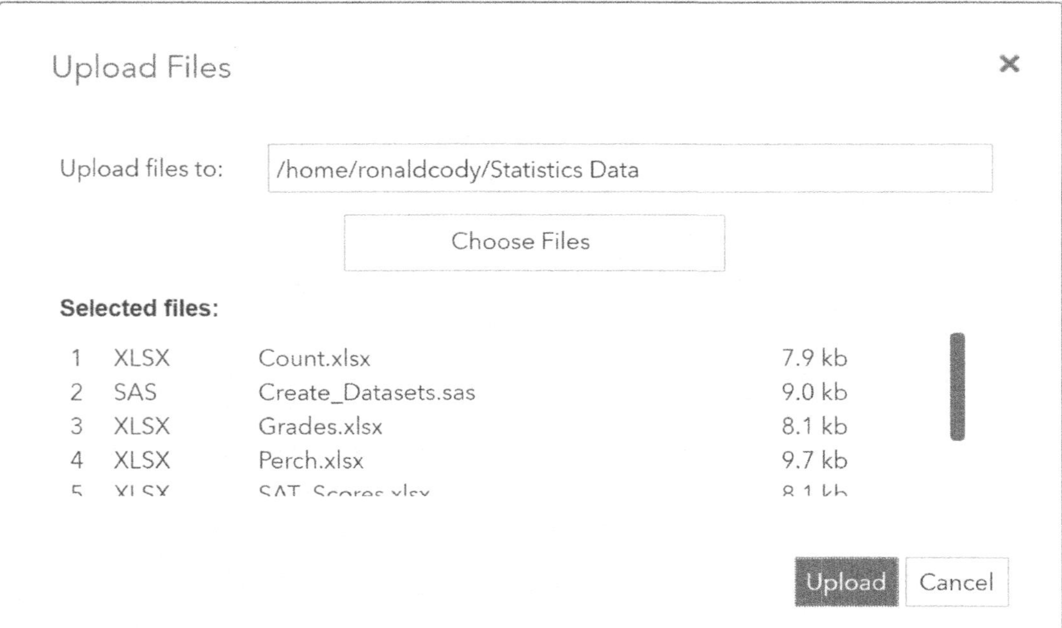

Figure 5.10: Getting Ready to Upload Selected Files

In Figure 5.11, you see all the selected files in the Statistics Data folder. Notice that the triangle to the left of Statistics Data points downward because you clicked it to reveal the individual files.

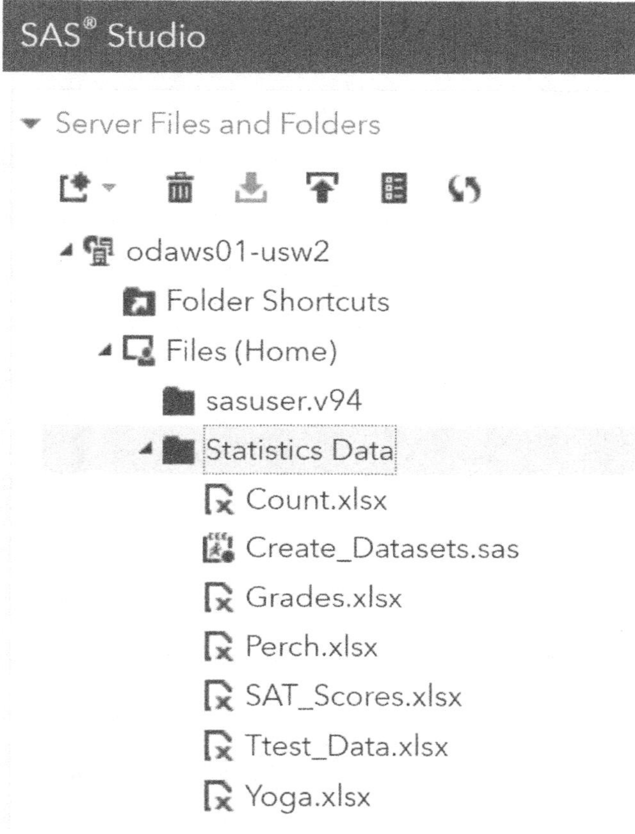

Figure 5.11: Files Uploaded to Statistics Data

You can now use SAS Studio to analyze any of these files or write a SAS program of your own using one or more of these files.

Before you can work with any data such as an Excel workbook or a CSV file, you need to convert it into a SAS data set. To accomplish this, you select the Import task that can be found under Tasks and Utilities task. (See Figure 5.12 below.)

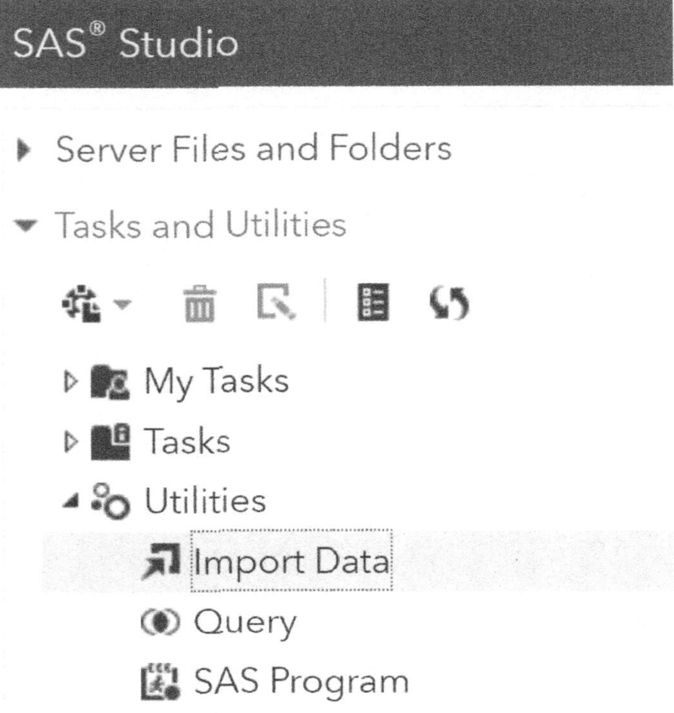

Figure 5.12: Selecting the Tasks and Utilities Task

Double click Import Data to begin converting the Excel workbook Grades.xlsx into a SAS data set.

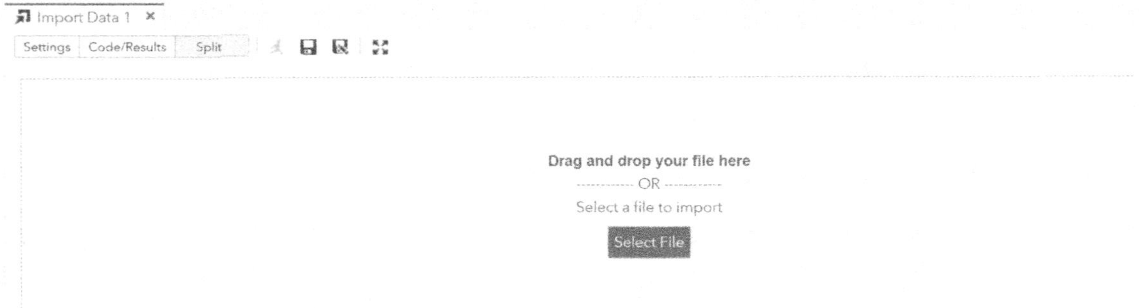

Figure 5.13: Selecting a File to Import

You have two ways to select which file you want to import. One is to click the Select File box on the right side of the screen—the other method is to click the Server Files and Folders tab in the navigation pane (on the left), find the file, and drag it to the drag and drop area.

Clicking Select File brings up a window where you can select a file to import. Here it is.

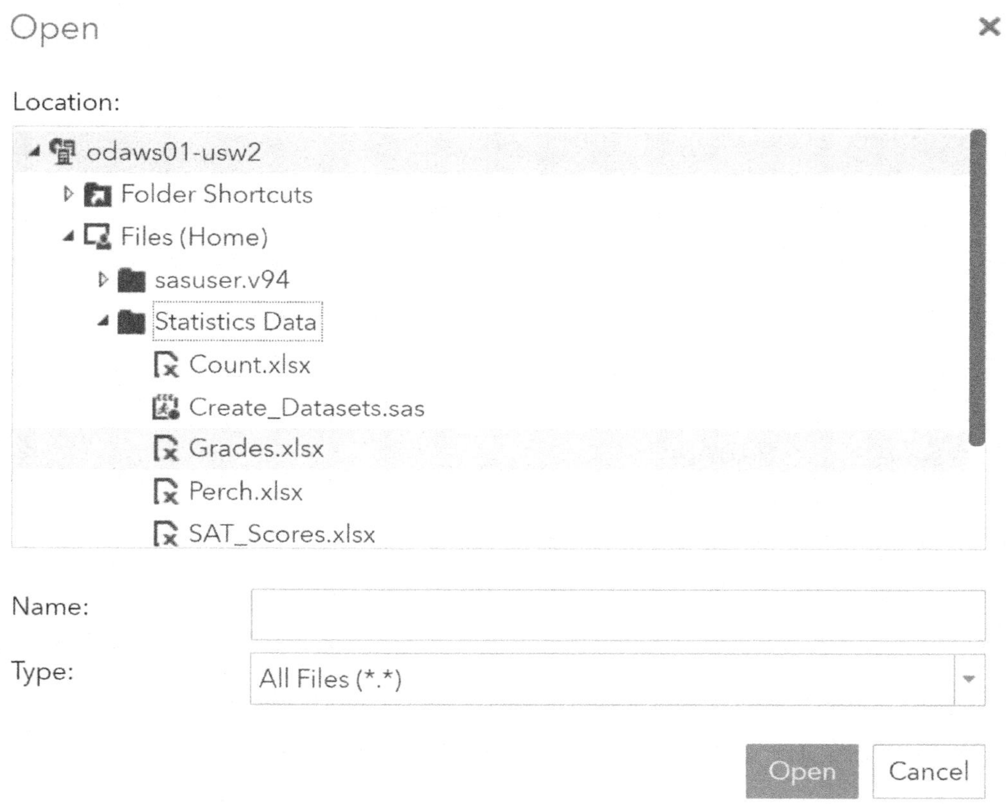

Open

Location:

odaws01-usw2
Folder Shortcuts
Files (Home)
sasuser.v94
Statistics Data
Count.xlsx
Create_Datasets.sas
Grades.xlsx
Perch.xlsx
SAT_Scores.xlsx

Name:

Type: All Files (*.*)

Open Cancel

Figure 5.14: Selecting the Grades Workbook

Click Grades.xlsx and then click Open. This brings up a split window showing options for the import on the top half and the SAS program that will be generated by these options in the bottom half.

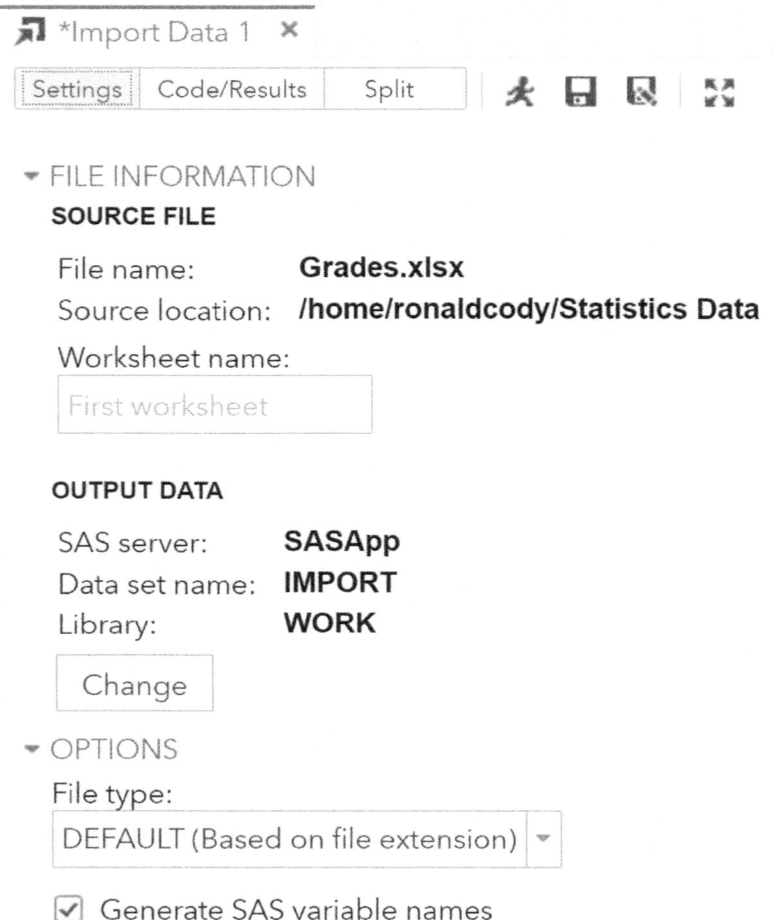

Figure 5.15: File Information

You might need to place your cursor at the bottom of this window to expand the window, enabling you see the selections for naming your output data set. There is an alternative. At the very top of the window are three tabs, labeled Settings, Code/Results, and Split. If you click the Settings tab, you do not have to expand the window manually.

Because you only have one worksheet, you do not have to enter a worksheet name. You probably want to change the name of the output (SAS) data set from the default name (IMPORT) to something more meaningful. You might also want to change the library where the SAS data set will be stored from the default WORK library. Clicking the Change button brings up a list of SAS libraries (below) and enables you to do both.

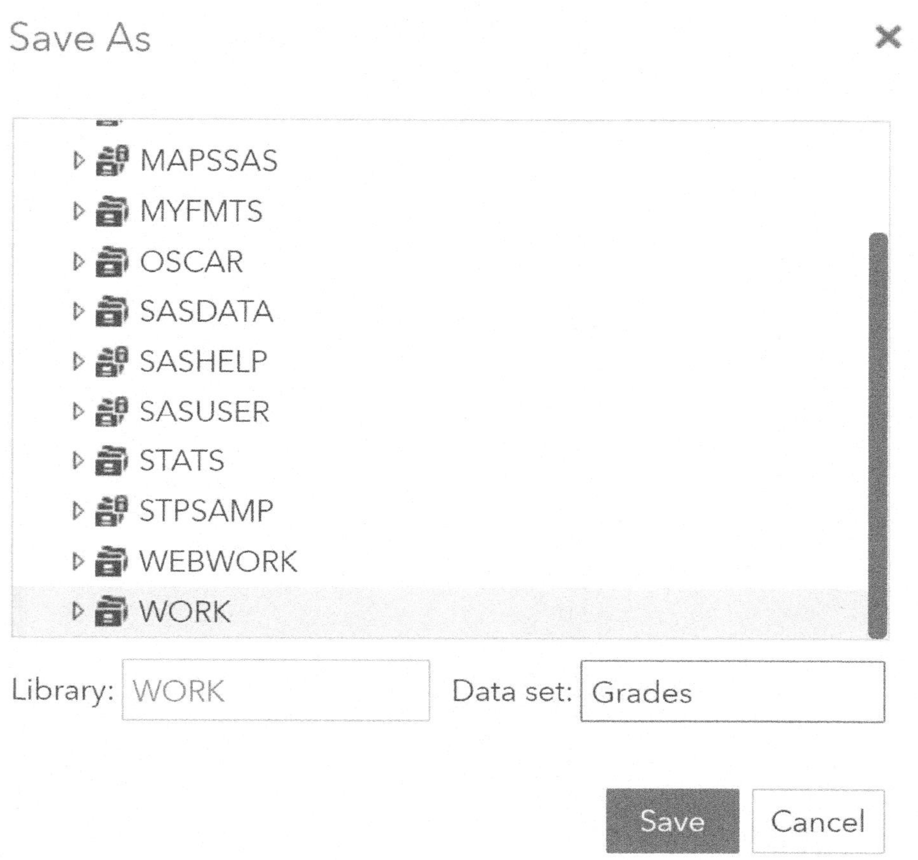

Figure 5.16: Selecting a File Name and a SAS Library

For this example, you are naming the file Grades and placing it in the default WORK library. This library is a temporary library whose contents are deleted when you exit your SAS session. Later in this chapter you will see how to create a permanent SAS library where your files will not be deleted when you close your SAS session.

Notice that you did not have to tell SAS Studio that you were importing an Excel workbook—it figured it out by the file extension (XLSX). SAS Studio will typically use the file extension to figure out how to import data. Because the first row of the spreadsheet contains variable names, leave the check on the Generate SAS variable names option. This tells the import utility to use the first row of the worksheet to generate variable names.

If you have a nonstandard file extension or if you prefer to manually select a file format, then you can use the drop-down list displayed in Figure 5.17 to instruct SAS how to convert your file.

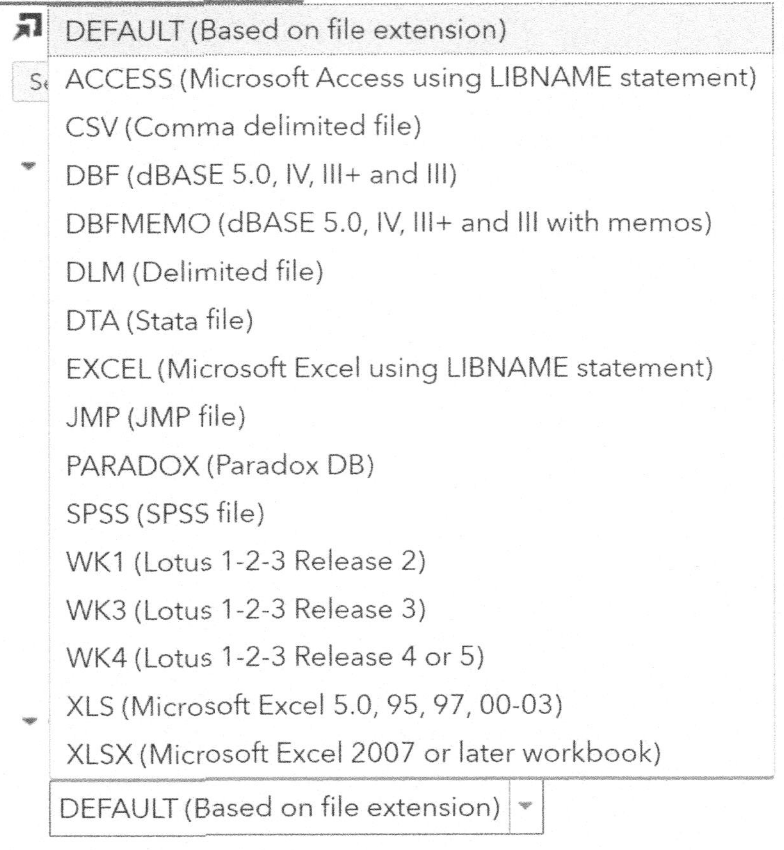

Figure 5.17: List of Supported File Types

When all is ready, click the Run icon (Figure 5.18).

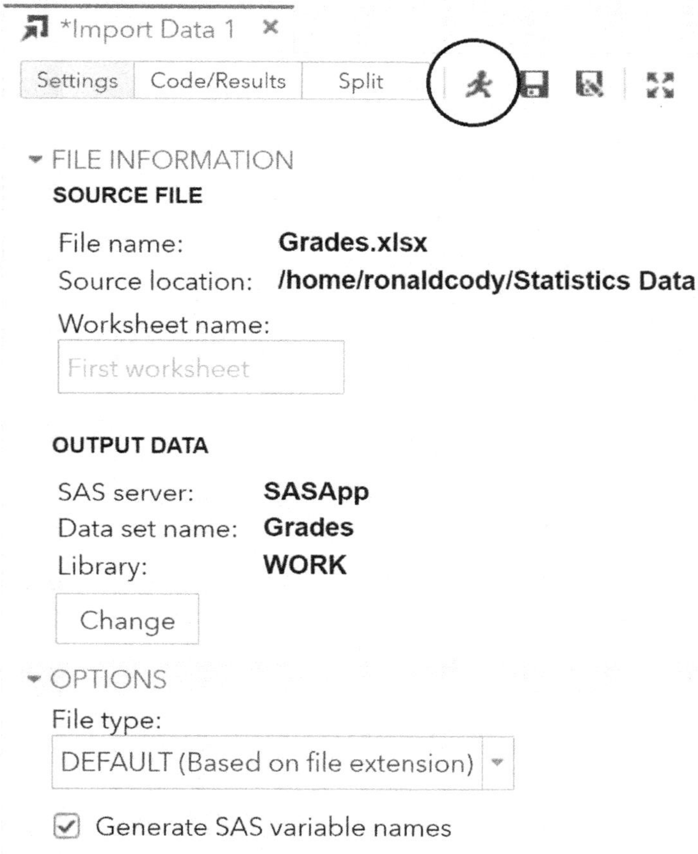

*Import Data 1

Settings | Code/Results | Split

FILE INFORMATION

SOURCE FILE

File name: **Grades.xlsx**

Source location: **/home/ronaldcody/Statistics Data**

Worksheet name:

First worksheet

OUTPUT DATA

SAS server: **SASApp**

Data set name: **Grades**

Library: **WORK**

Change

OPTIONS

File type:

DEFAULT (Based on file extension)

Generate SAS variable names

Figure 5.18: The Run Icon

SAS Studio will now list information about the Grades data set. Selections of the output are shown in Figure 5.19.

Data Set Name	WORK.GRADES	**Observations**	3
Member Type	DATA	**Variables**	8
Engine	V9	**Indexes**	0
Created	02/27/2021 10:05:30	**Observation Length**	72
Last Modified	02/27/2021 10:05:30	**Deleted Observations**	0
Protection		**Compressed**	NO
Data Set Type		**Sorted**	NO
Label			
Data Representation	SOLARIS_X86_64, LINUX_X86_64, ALPHA_TRU64, LINUX_IA64		
Encoding	utf-8 Unicode (UTF-8)		

Alphabetic List of Variables and Attributes						
#	Variable	Type	Len	Format	Informat	Label
8	Final	Num	8	BEST.		Final
2	ID	Num	8	BEST.		ID
5	Midterm	Num	8	BEST.		Midterm
1	Name	Char	10	$10.	$10.	Name
3	Quiz1	Num	8	BEST.		Quiz1
4	Quiz2	Num	8	BEST.		Quiz2
6	Quiz3	Num	8	BEST.		Quiz3
7	Quiz4	Num	8	BEST.		Quiz4

Figure 5.19: Edited Output from the Import Task

At the bottom of this output, you see a list of the variable names, whether they are stored as numeric or character, along with some other information that we don't need at this time. Notice that the import utility correctly read Name as character and the other variables as numeric.

Listing the SAS Data Set

A quick way to see a listing of the Grades data set is to select the Libraries tab in the navigation pane and select My Libraries.

Figure 5.20: The Libraries Tab

Expand the Work library and double-click Grades. It looks like Figure 5.21.

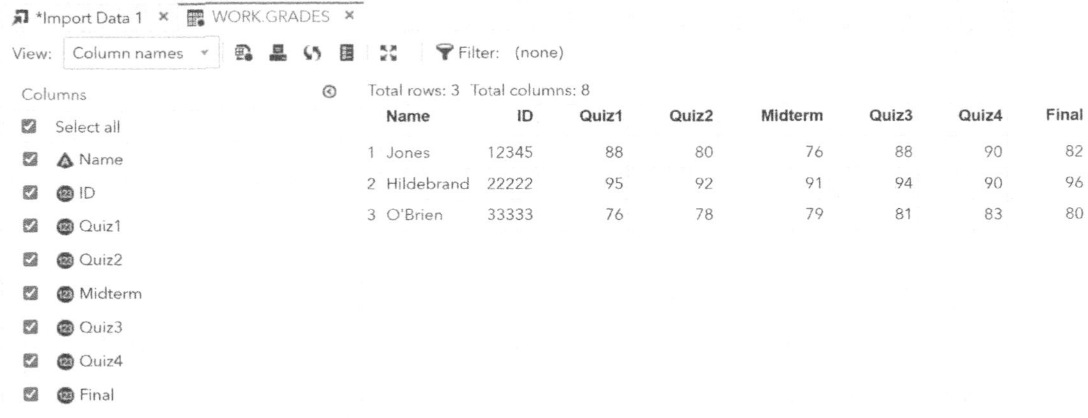

	Name	ID	Quiz1	Quiz2	Midterm	Quiz3	Quiz4	Final
1	Jones	12345	88	80	76	88	90	82
2	Hildebrand	22222	95	92	91	94	90	96
3	O'Brien	33333	76	78	79	81	83	80

Figure 5.21: Data Set Grades in the Work Library

You can use your mouse to scroll to the right to see the rest of the table.

> In this book, as well as in SAS Studio, you will see the terms SAS data set and table used interchangeably as well as these other equivalent terms: variables are also called columns and observations are called rows.

To create a nicer looking report, click the Tasks and Utilities tab of the navigation pane and select Data followed by List Data, as in Figure 5.22.

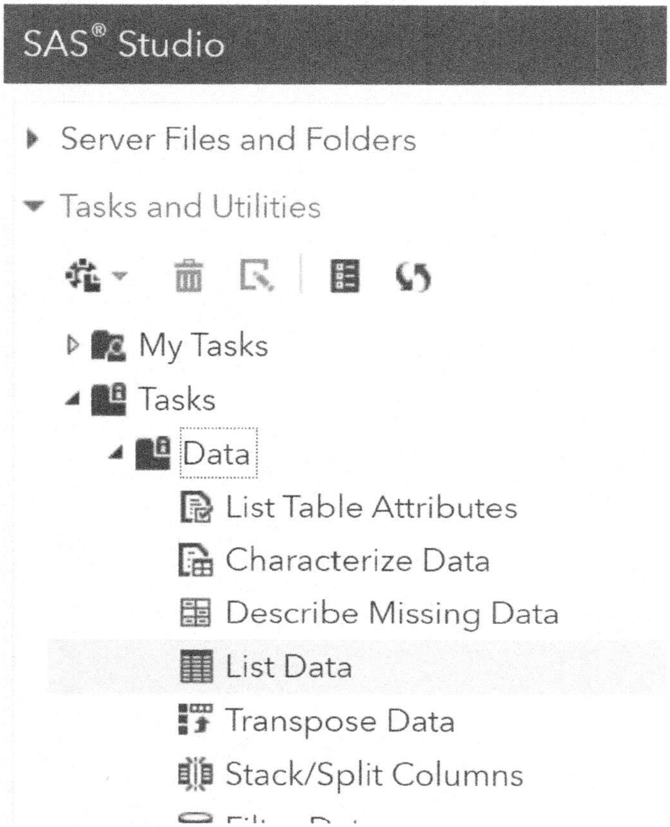

Figure 5.22: The List Data Task

Double-click List Data. The screen that appears resembles Figure 5.23.

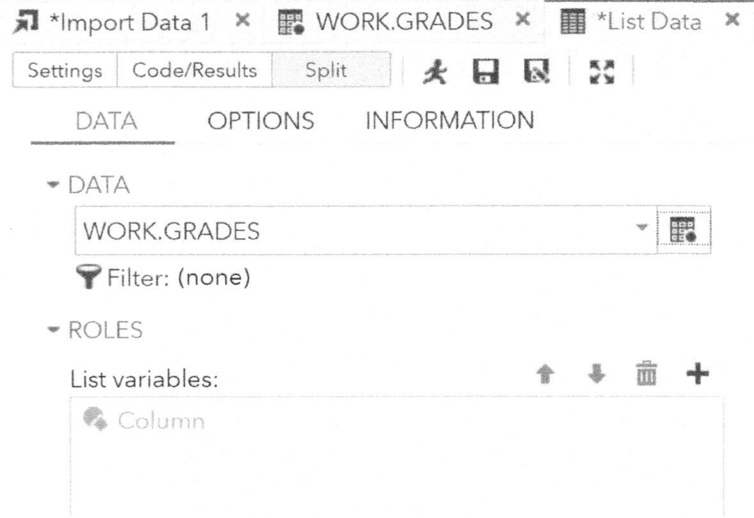

Figure 5.23: List Data Task

Click the icon at the far right side of the box labeled DATA (circled in the figure below).

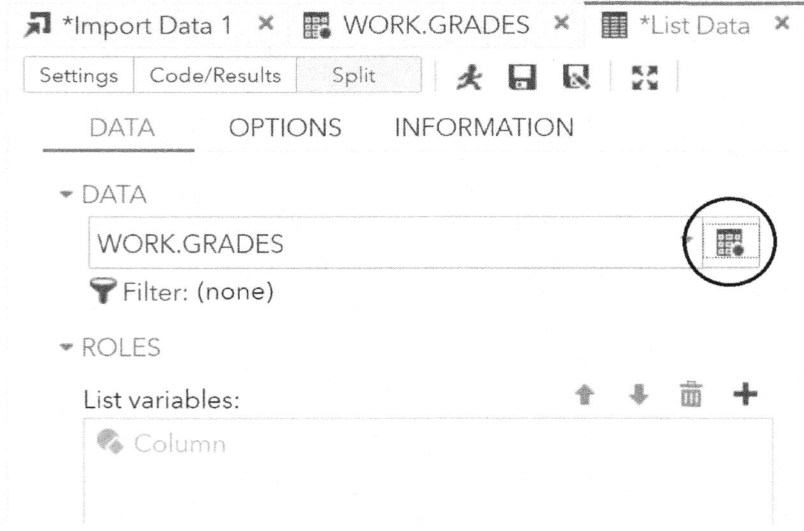

Figure 5.24: Select Table

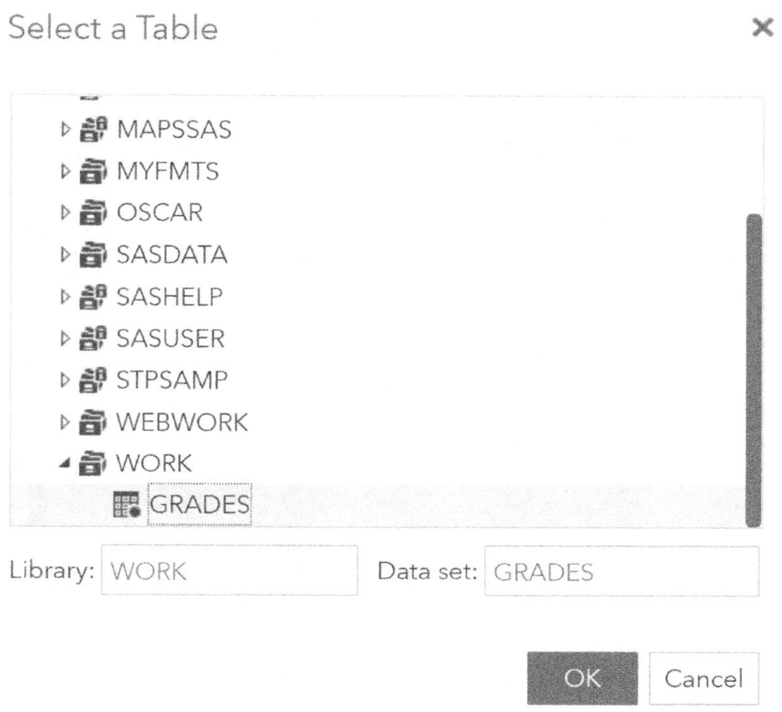

Select a Table

MAPSSAS
MYFMTS
OSCAR
SASDATA
SASHELP
SASUSER
STPSAMP
WEBWORK
WORK
GRADES

Library: WORK
Data set: GRADES

OK Cancel

Figure 5.25: Select Grades in the WORK Library

Click OK. Your next task is to select which variables you want in the listing. Under ROLES, you see four boxes. At this time, you are only interested in the first two: List variables and Group analysis by. Click the plus sign next to List Variables. A list of variables from the Grades data set will show up (Figure 5.26).

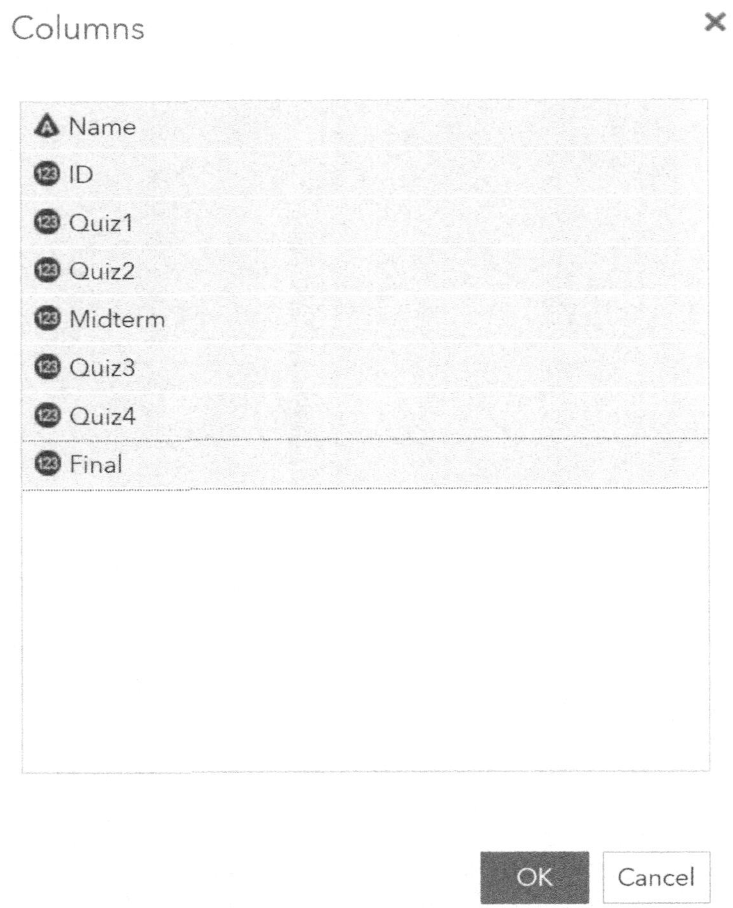

Columns

Name
ID
Quiz1
Quiz2
Midterm
Quiz3
Quiz4
Final

OK Cancel

Figure 5.26: Select Variables

Select the variables that you want to see in your listing. If you want to see all of the variables, first click Name. Then hold down the Shift key and click Final. Click OK. Your screen now should look like Figure 5.27.

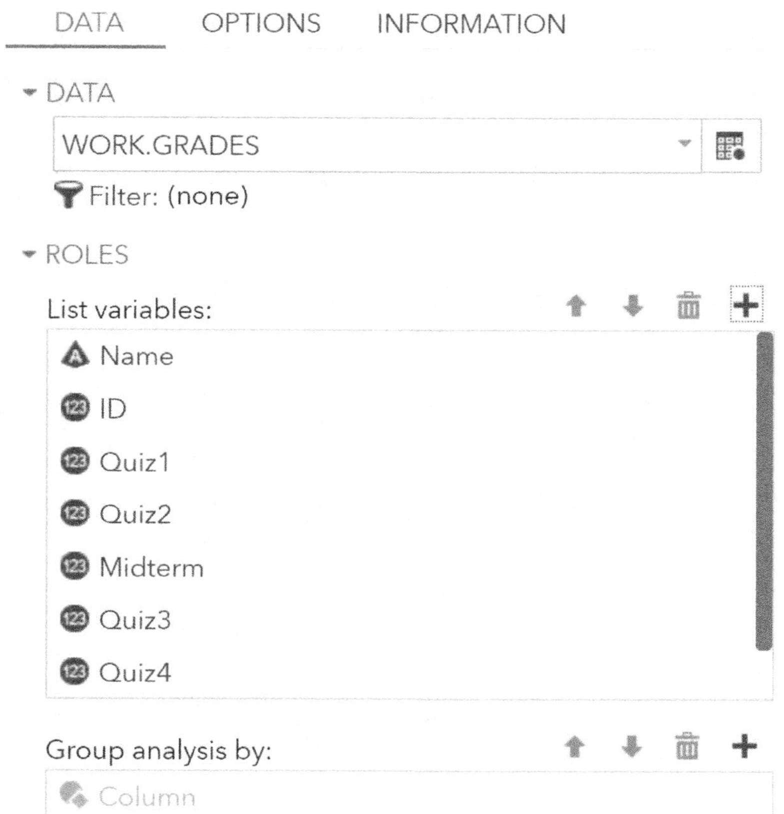

Figure 5.27: Tables and Variables Selected

Click the Run icon and the listing of the Grades data set appears as shown in Figure 5.28.

List Data for WORK.GRADES

Obs	Name	ID	Quiz1	Quiz2	Midterm	Quiz3	Quiz4	Final
1	Jones	12345	88	80	76	88	90	82
2	Hildebrand	22222	95	92	91	94	90	96
3	O'Brien	33333	76	78	79	81	83	80

Figure 5.28: Listing of Data Set Grades

Importing Data from a CSV File

CSV (comma-separated values) files are a popular format for external data files. As the name implies, CSV files use commas as data delimiters. Many websites enable you to download data as CSV files. As with Excel workbooks, your CSV file might or might not contain variable names at the beginning of the file. If the file does contain variable names, be sure the Generate SAS variable names options box is checked; if not, uncheck this option.

For example, take a look at the CSV file called Grades.csv in Figure 5.29.

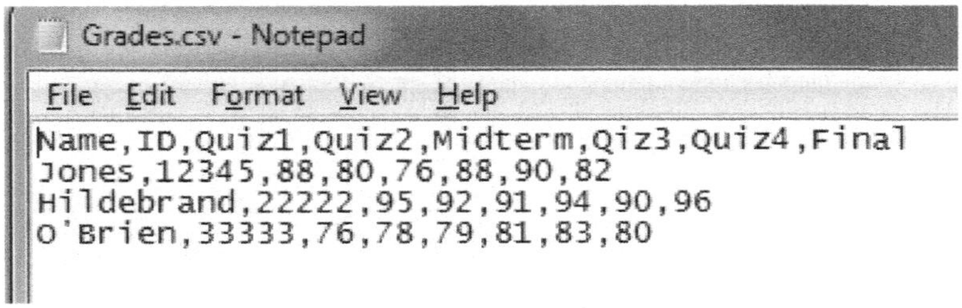

Grades.csv - Notepad

File Edit Format View Help

```
Name,ID,Quiz1,Quiz2,Midterm,Qiz3,Quiz4,Final
Jones,12345,88,80,76,88,90,82
Hildebrand,22222,95,92,91,94,90,96
O'Brien,33333,76,78,79,81,83,80
```

Figure 5.29: CSV File Grades.csv

This CSV file contains the same data as the Excel workbook Grades.xlsx. Notice that variable names are included in the file. You can import this file and create a SAS data set, using the same steps that you used to import the Excel workbook. The import facility will automatically use the correct code to import this data file because of the CSV file extension. The resulting SAS data set is identical to the one shown in Figure 5.28.

Conclusion

You have seen how easy it is to import data in a variety of formats and create SAS data sets. Even experienced programmers (at least the ones this author knows) would prefer to use the import data utility to convert external data to SAS data sets rather than writing their own code.

6.

Descriptive Statistics – Univariate Analysis

Introduction

Before you begin any statistical test, you should spend some time "getting to know your data." This chapter describes ways to examine both continuous and categorical data using a variety of techniques, including descriptive statistical measures such as means and standard deviations as well as graphical techniques such as histograms and box plots. The term **univariate** describes single variables and not the relationship between variables, which is a topic discussed in several of the later chapters in this book (such as correlation and regression).

This step is so important because understanding your data is necessary when you are choosing appropriate statistical tests to perform. Also, describing your data, especially using graphical techniques, is one way to spot possible errors in your data.

Generating Descriptive Statistics for Continuous Variables

Let's use the Cars data set, which is located in the SASHELP library, to demonstrate how to produce descriptive statistics for continuous and categorical variables. Start by clicking the triangle to the left of Tasks (or double-click anywhere on the tab) to expand the list of subtasks available (Figure 6.1). Next, either click the triangle to the left of Statistics or double click anywhere on the Statistics task.

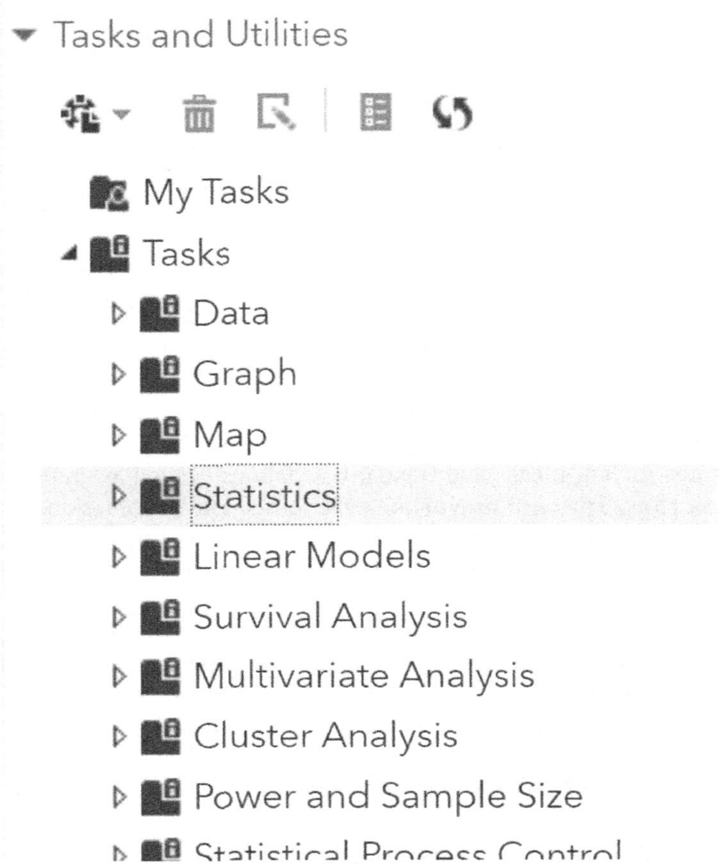

Figure 6.1: Selecting Summary Statistics from the Statistics Menu

This brings up the DATA, OPTIONS, OUTPUT, and INFORMATION tabs as shown in Figure 6.2.

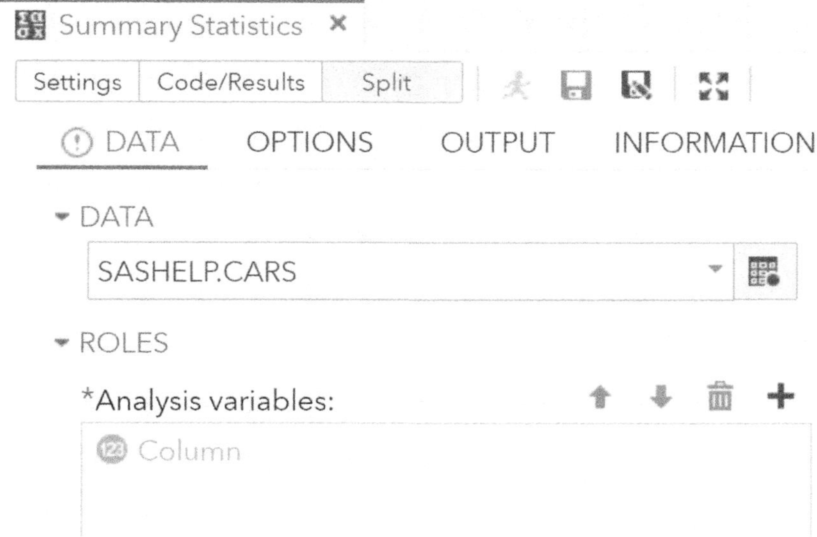

Figure 6.2: DATA Tab for Summary Statistics

As with almost every statistical task, the first tab you see is the DATA tab. It is here where you select the SAS data set that contains your data and select which variables you need for various roles. Because you want to analyze data from the Cars data set in the SASHELP library, you click the Select a Table icon, choose the SASHELP library and the Cars data set. This data set contains variables such as the price of the car (MSRP), the weight, type of car, and so on.

The next step is to select variables to analyze. Click the plus sign in the Roles section of the pane to bring up a list of variables in the Cars data set. You can select variables in two ways. One is to hold down the Ctrl key and left-click each of the variables that you want to select. The other method is to click one variable, hold down the Shift key, and then click a variable farther down in the list. All the variables from the first to the last will be selected. You can even combine these two methods to select variables. In the following example, the variables of interest are not next to each other in the list. Therefore, you can hold down the Ctrl key and click each of the variables MSRP, Horsepower, and Weight. It looks like this:

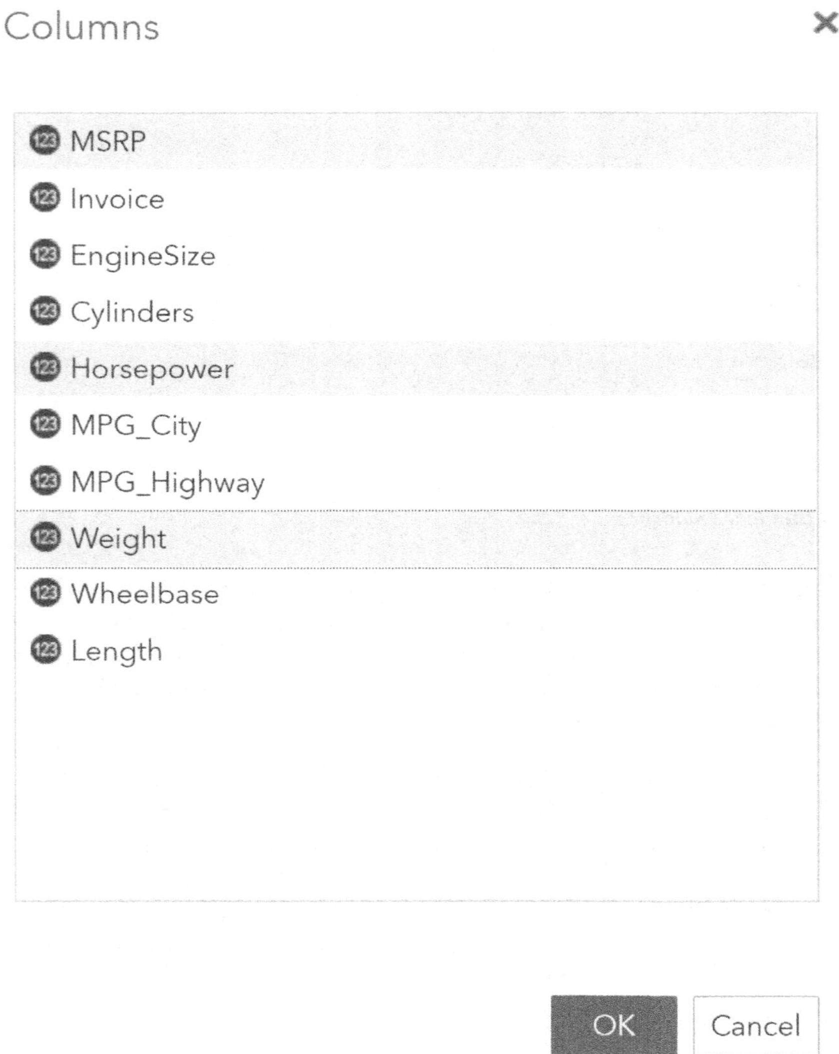

Figure 6.3: Selecting Variables for Analysis

Once you click OK, you can click the OPTIONS tab to select or deselect statistics and plots that you want to generate (see Figure 6.4).

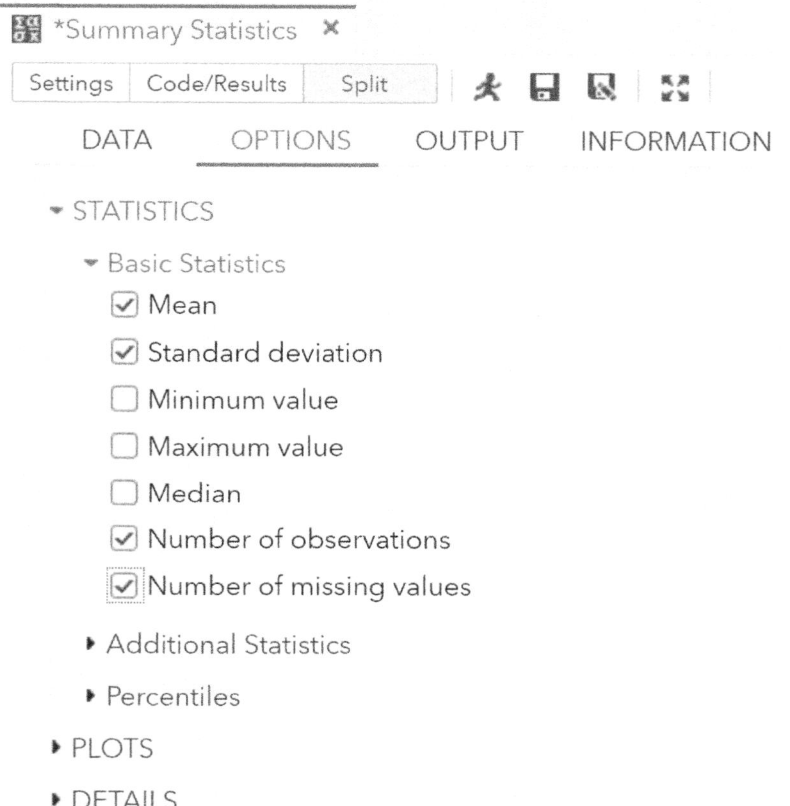

Figure 6.4: OPTIONS Tab

Notice that many of the statistics boxes are already checked. You can select additional statistics or click a box to deselect a statistic that has already been selected. In this example, the Number of missing values and a request for the median have been added to the default list and the two options Minimum and Maximum value have been deselected. It is very useful to see both the number of nonmissing observations along with the number of observations with missing values.

One other useful statistic is the 95% confidence interval (95% CI) for the mean. The 95% confidence interval for the mean is useful in determining how accurately your sample mean estimates the mean of the population from which you drew your sample.

To request this statistic, click the triangle to the left of the heading Additional Statistics. This reveals a further set of choices as shown in Figure 6.5.

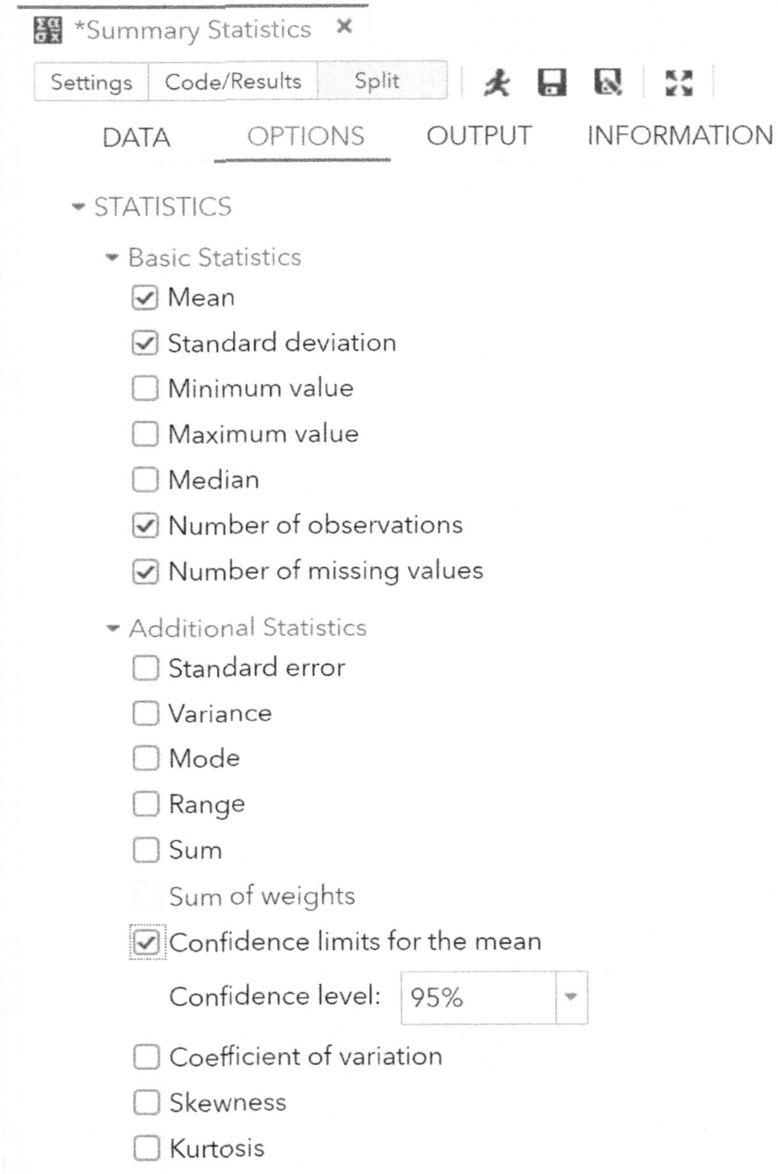

*Summary Statistics ×

Settings | Code/Results | Split

DATA | OPTIONS | OUTPUT | INFORMATION

▾ STATISTICS

▾ Basic Statistics

- [x] Mean
- [x] Standard deviation
- [ ] Minimum value
- [ ] Maximum value
- [ ] Median
- [x] Number of observations
- [x] Number of missing values

▾ Additional Statistics

- [ ] Standard error
- [ ] Variance
- [ ] Mode
- [ ] Range
- [ ] Sum
- [ ] Sum of weights
- [x] Confidence limits for the mean

  Confidence level: 95%
- [ ] Coefficient of variation
- [ ] Skewness
- [ ] Kurtosis

Figure 6.5: Additional Statistics

When you check the box for Confidence limits for the mean, the option below, labeled Confidence level, displays the default value of 95% for the confidence level. You can select other intervals, but 95% is the one most commonly used.

Finally, you can also select plots using the OPTIONS tab. Here you are requesting a histogram and box plot for the selected variables (Figure 6.6 below).

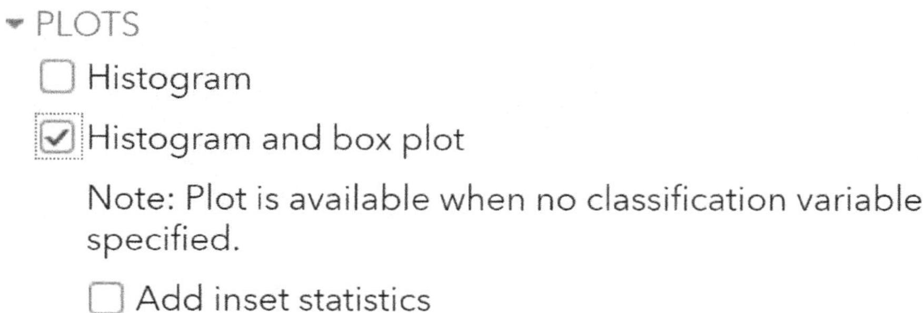

Figure 6.6: Requesting a Histogram and Box Plot

You are ready to click the Run icon.

The first section of output shows basic statistics for the selected variables.

Variable	Label	Mean	Std Dev	N	N Miss	Lower 95% CL for Mean	Upper 95% CL for Mean
MSRP		32774.86	19431.72	428	0	30928.69	34621.02
Horsepower		215.8855140	71.8360316	428	0	209.0605358	222.7104922
Weight	Weight (LBS)	3577.95	758.9832146	428	0	3505.84	3650.06

Figure 6.7: Descriptive Statistics for Selected Variables

You see the mean, standard deviation, median, the number of nonmissing values, and the number of missing values for each of the analysis variables. The last two columns in the table represent the lower and upper 95% confidence limits for the mean.

The next section of output consists of a histogram and, directly below it, the box-plot for each variable. To save space, only two histograms, one for MSRP and one for Horsepower, are shown in the two figures that follow.

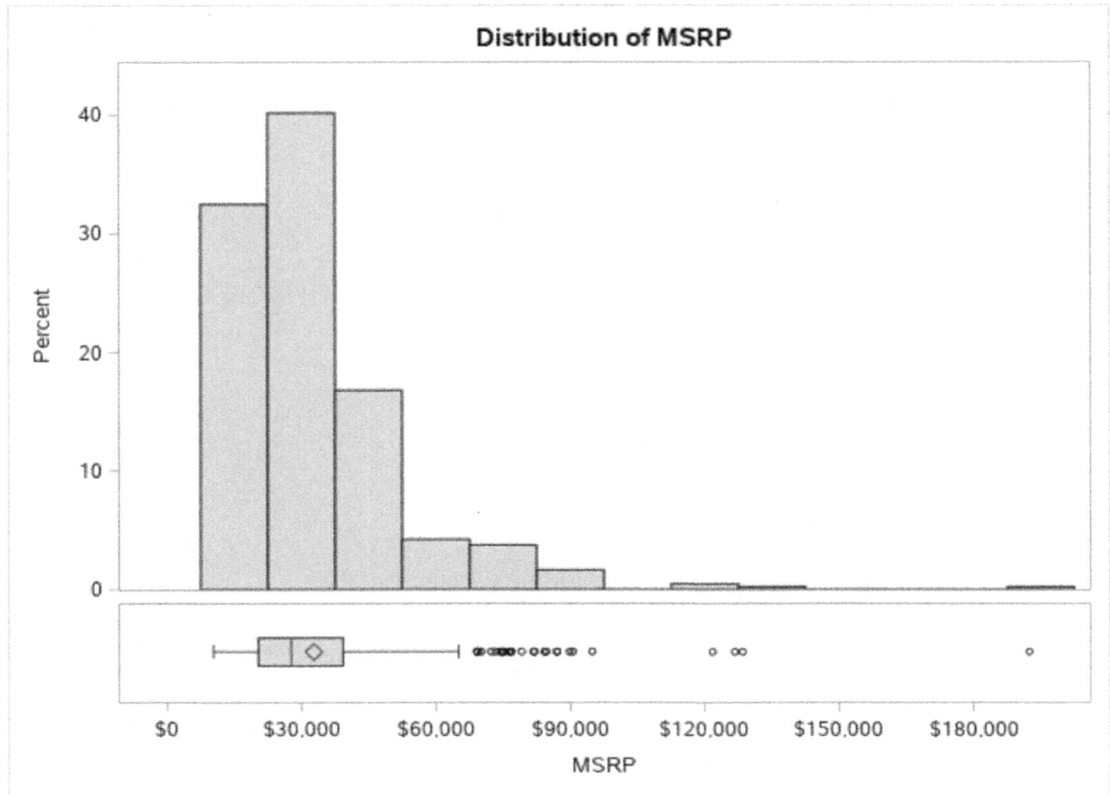

Figure 6.8: Histogram and Box Plot for MSRP

You can see that the histogram for MSRP has some very large values on the right side of the distribution (probably because of some very expensive luxury cars). This distribution is described by statisticians as being **skewed** to the right. If there are extreme values on the left of the distribution, it is said to be skewed to the left—the terms right and left indicate which side of the distribution contains extreme values.

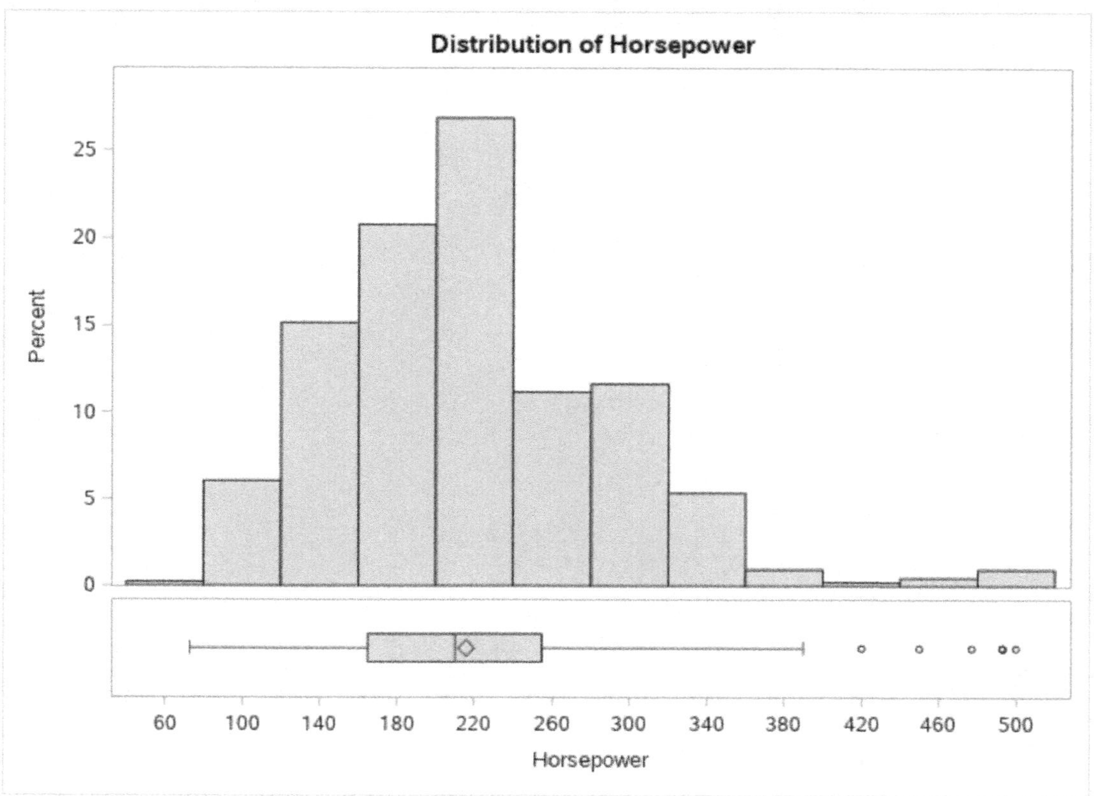

Figure 6.9: Histogram and Box Plot for Horsepower

A detailed discussion of box plots is presented later in this chapter. For now, we will concentrate on the histogram. The histogram for Horsepower is also positively skewed, easily seen by the long tail on the right side of the histogram.

Investigating the Distribution of Horsepower

Let's use the variable Horsepower to demonstrate how to further investigate the distribution of a continuous variable. One way to do this is to select Distribution Analysis from the list of Statistics tasks. Make sure that the Cars data set is selected on the DATA tab and Horsepower is selected as the analysis variable. Click the OPTIONS tab to bring up the following menu:

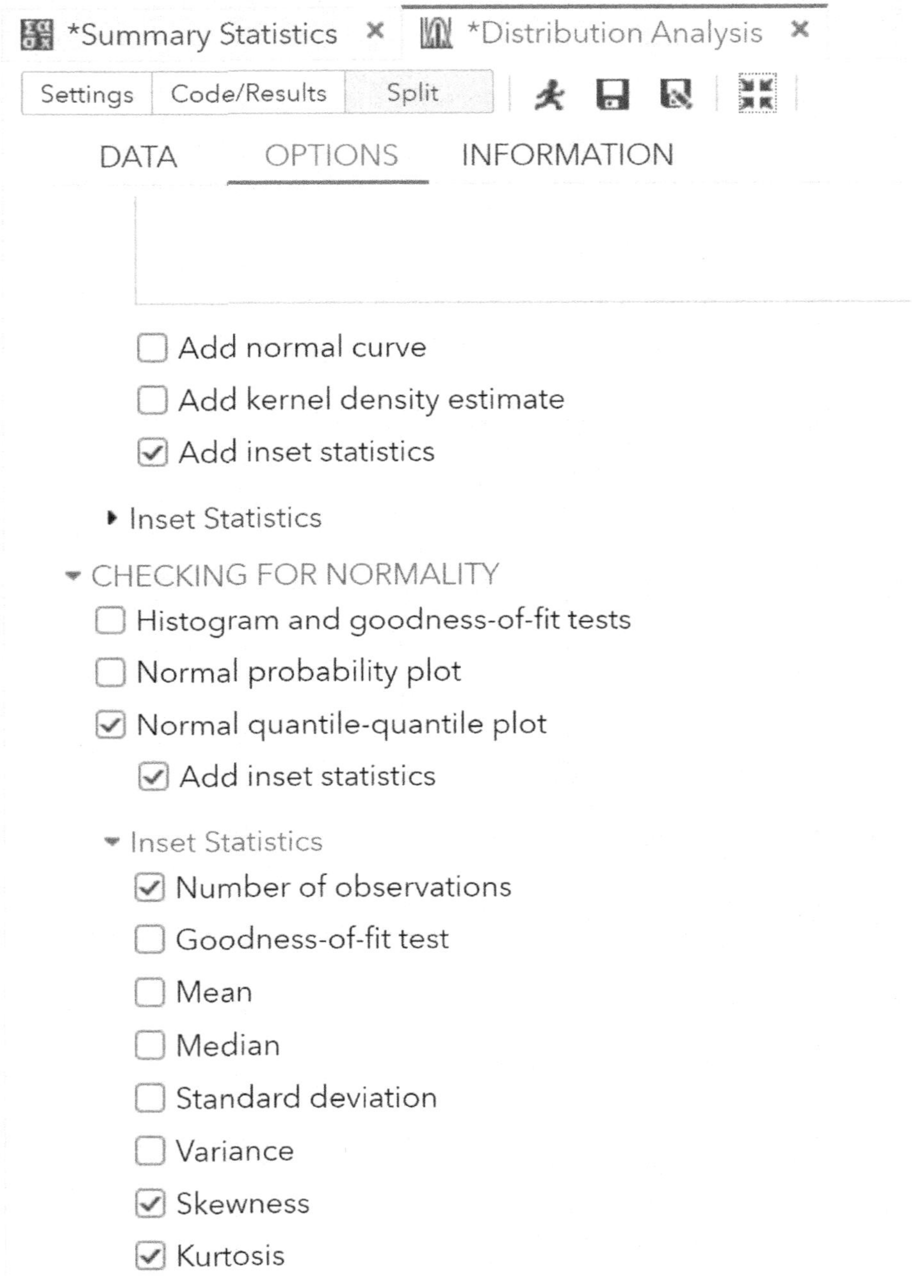

*Summary Statistics ✕ *Distribution Analysis ✕

Settings | Code/Results | Split

DATA OPTIONS INFORMATION

- [ ] Add normal curve
- [ ] Add kernel density estimate
- [x] Add inset statistics

▸ Inset Statistics

▾ CHECKING FOR NORMALITY

- [ ] Histogram and goodness-of-fit tests
- [ ] Normal probability plot
- [x] Normal quantile-quantile plot
  - [x] Add inset statistics

▾ Inset Statistics

- [x] Number of observations
- [ ] Goodness-of-fit test
- [ ] Mean
- [ ] Median
- [ ] Standard deviation
- [ ] Variance
- [x] Skewness
- [x] Kurtosis

Figure 6.10: Options for the Distribution Analysis Tab

Because you have already produced a histogram from the Summary Statistics tab, you first want to deselect the box next to Histogram. Next, you have a choice of options for checking for normality. In this example, you are requesting a Normal Quantile-Quantile (Q-Q) plot with added inset statistics (values placed in a box on the plot). A Q-Q plot displays the quantiles of one distribution on the X axis and the quantiles of another distribution on the Y axis. A **quantile** is the proportion or percent of a distribution that falls below a given value. For example, 25% of the data values will fall below the 25th percentile. The Q-Q plot produced by SAS displays the quantiles of a theoretical distribution (in this case, a normal distribution) on the X axis and the actual quantile for your sample distribution on the Y axis. If you have normally distributed data, the Q-Q-plot will fall along a straight line.

Two popular statistics that quantify deviations from normality, Skewness and Kurtosis, are selected to be displayed in an inset box on the Q-Q-plot. We will discuss these two terms shortly. Clicking the Run icon produces the following plot:

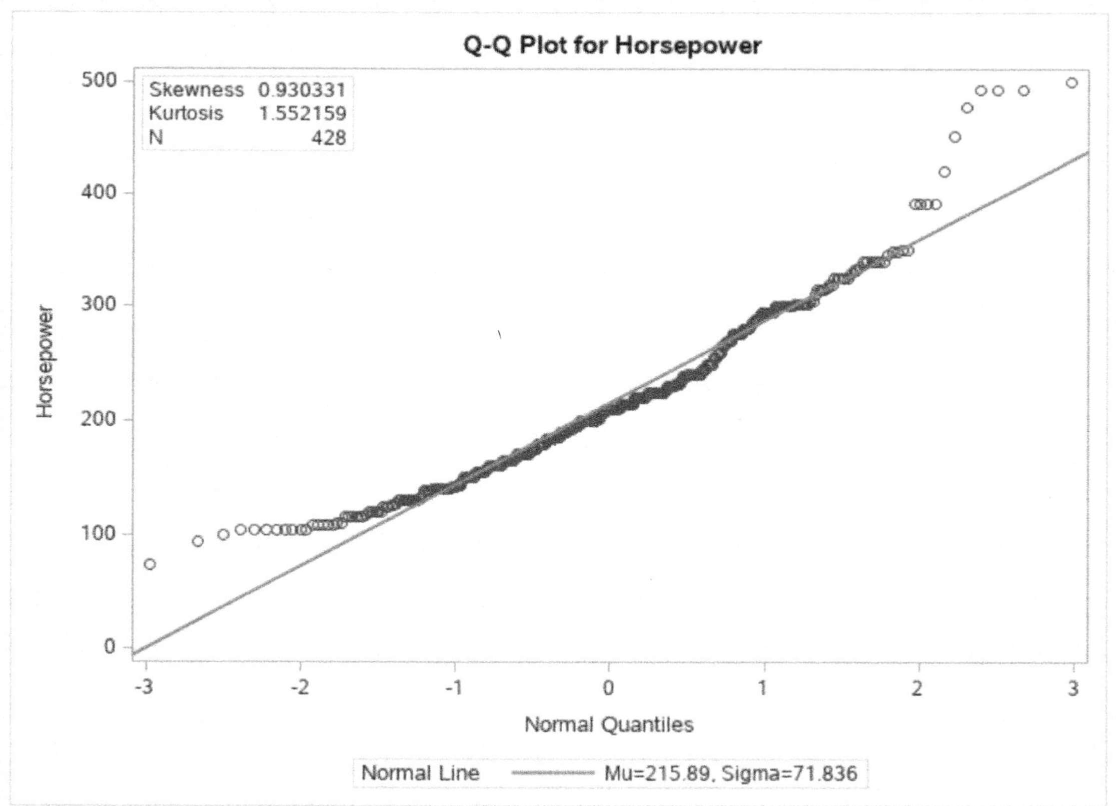

Figure 6.11: Q-Q-Plot for Horsepower

The straight line on the plot represents a normal distribution with the same mean and standard deviation as the variable Horsepower. The circles on the plot represent values of horsepower from your sample data. At the bottom of the Q-Q plot, you see that the theoretical normal distribution has a mean (Mu) equal to 215.89 and a standard deviation (Sigma) equal to 71.836.

To help you understand this Q-Q plot, look at the right side of the plot. The circles above the straight line on this part of the plot indicate that your sample data includes values of horsepower that are higher (more extreme) than you would expect if horsepower values were normally distributed. This confirms the strong positive skewness that you saw in the histogram.

Values for Skewness and Kurtosis close to zero result from distributions that are close to normal. Positive values for skewness, as in this plot, indicate a positively skewed distribution (extreme values in the right tail). Positive values for **kurtosis** (as in this example) indicate both that the distribution is too peaked (**leptokurtic**) and that the tails (left and right side of the distribution) contain more data values than a normal distribution. Negative values for kurtosis indicate that the distribution is too flat (**platykurtic**) and that there are too few data values in the tails of the distribution.

When it is time to run statistical tests on horsepower and various categorical variables of interest, you might be concerned that the distribution for horsepower deviates quite noticeably from a normal distribution. Because the sample size of the Cars data set is relatively large (428 observations), you might feel comfortable in running parametric tests such as *t* tests and ANOVA. **Parametric** tests rely on the data values being distributed in a specific way, such as a normal distribution. **Nonparametric** methods are often described as distribution-free methods. Those types of decisions will be explored in later chapters that discuss inferential statistics.

Adding a Classification Variable in the Summary Statistics Tab

Suppose you want to see if horsepower is related to the number of cylinders? It would be reasonable to assume that vehicles with more cylinders would have more power. Before you begin this investigation, it would be a good idea to investigate the Cylinders variable. The next section describes how to do this.

Computing Frequencies for Categorical Variables

The first step is to select One-Way Frequencies from the Statistics menu, as shown in Figure 6.12.

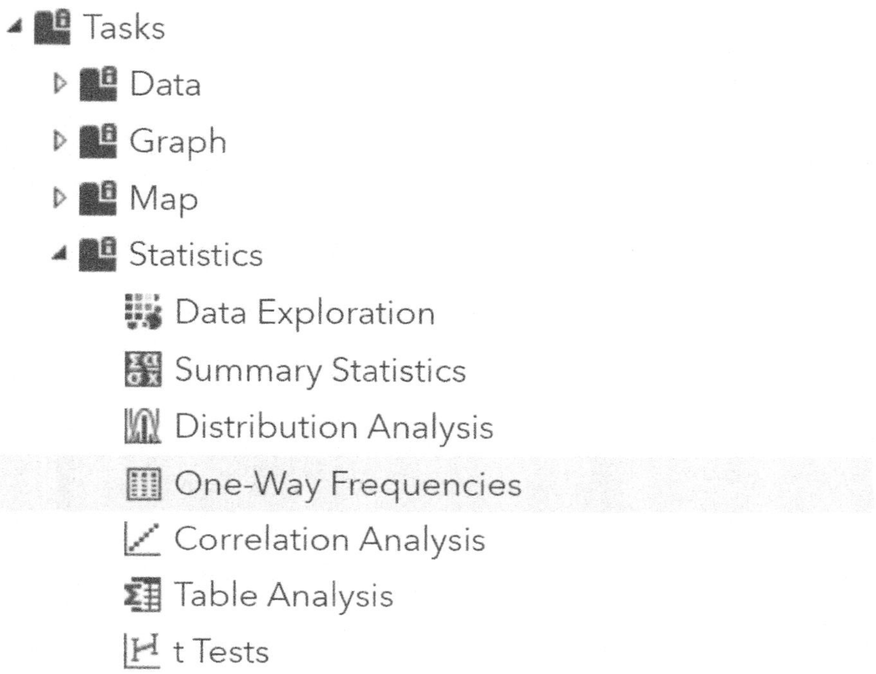

Tasks
Data
Graph
Map
Statistics
Data Exploration
Summary Statistics
Distribution Analysis
One-Way Frequencies
Correlation Analysis
Table Analysis
t Tests

Figure 6.12: Select One-Way Frequencies from the Statistics Task

Next, be sure the Cars data set in the SASHELP library is selected. Under the Roles tab, click the plus (+) sign and select the variable Cylinders.

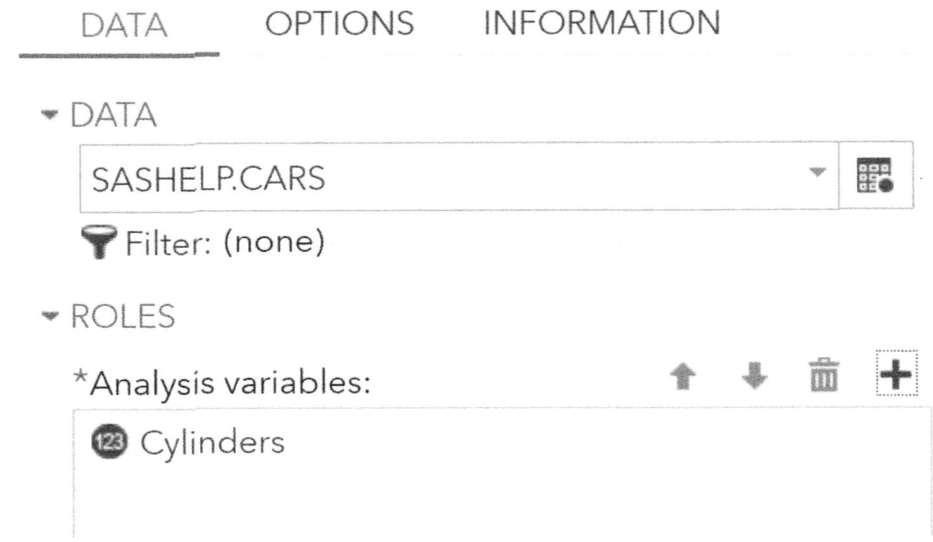

Figure 6.13: Select Your Analysis Variable(s)

Click the Run icon to obtain the output displayed in Figure 6.14.

Cylinders	Frequency	Percent	Cumulative Frequency	Cumulative Percent
3	1	0.23	1	0.23
4	136	31.92	137	32.16
5	7	1.64	144	33.80
6	190	44.60	334	78.40
8	87	20.42	421	98.83
10	2	0.47	423	99.30
12	3	0.70	426	100.00
Frequency Missing = 2				

Figure 6.14: Frequencies for the Variable Cylinders

Who knew there were three- and five-cylinder cars? Because some of these categories contain so few observations, let's restrict our comparison of horsepower to four- and six-cylinder cars. To accomplish this, you need to create a filter, which is the subject of the next section.

Creating a Filter Within a Task

You can select which rows in a table to include in an analysis. It's easy to do. Go back to the Distribution Analysis task and on the DATA tab, click the Filter icon. (See figure below.)

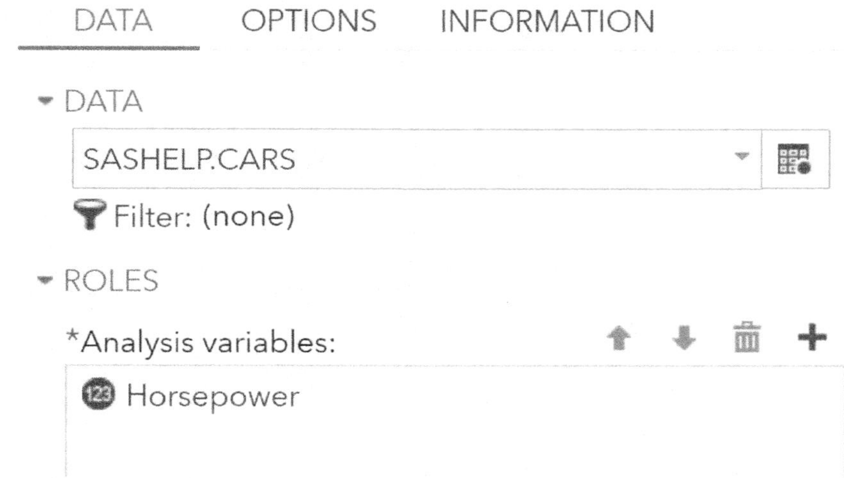

Figure 6.15: Creating a Filter

This brings up the filter box as shown next.

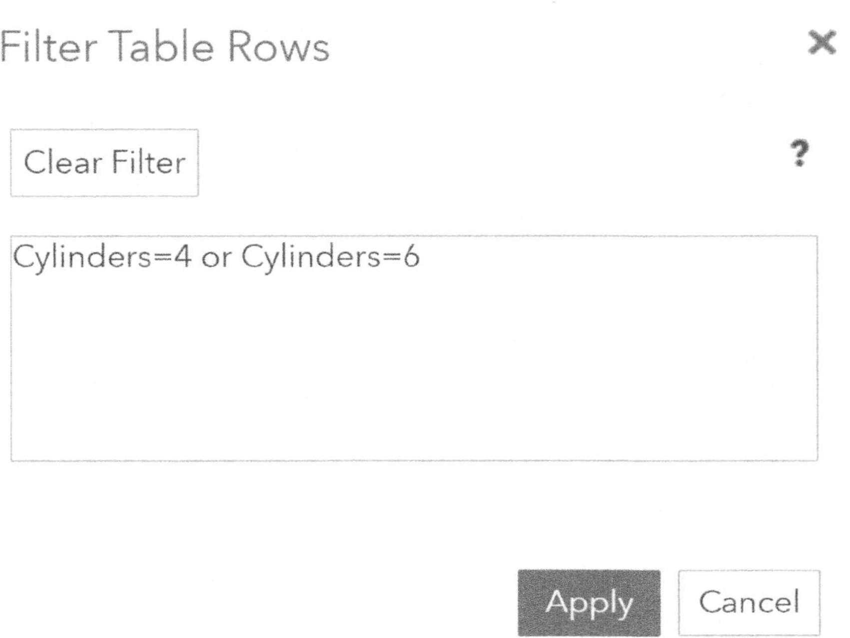

Figure 6.16: Selecting Cars with Four or Six Cylinders

If you are familiar with how to use a WHERE clause in SQL, you already know how to create a filter—simply write the contents of the WHERE clause, leaving out the word WHERE. If you are unfamiliar with WHERE clauses, it's quite simple. In the example shown here, you are asking if the variable Cylinders is equal to 4 or 6. It is very important not to write this expression as:

```
Cylinders = 4 or 6
```

This expression makes sense if you are speaking to another person, but not as a computer statement. You have to explicitly repeat the name of the variable for each part of this expression as shown in Figure 6.16.

> Because SAS treats any numeric value other than zero or missing as "true," the expression above would evaluate each part of the OR operator. One part is Cylinders = 4. The other part is 6. Because 6 is not zero or missing it is evaluated as "true." If one part of an OR expression is true, the entire expression is true. Strange as it sounds, the expression above will not cause an error message to be printed, and all values of the variable Cylinders will be included in the analysis.

Other examples of filter expressions are shown below:

Filter Expression	Filter Syntax
Horsepower greater than 150	Horsepower > 150
Horsepower between 100 and 200	Horsepower between 100 and 200
Horsepower is not missing	Horsepower is not missing alternative: Horsepower is not null
Gender is M or F	Gender = ‘M’ or Gender = ‘F’
Cylinders is not equal to 3 or 5	Cylinders ne 3 and Cylinders ne 5

Notice the quotation marks around the M and F in the Gender example. You need quotation marks (either single or double) for all character values. Whenever SAS Studio shows you a list of variables to be selected, notice the symbols 123 or A before the variable names. The symbol 123 indicates a numeric variable, and the A in a triangle represents a character (Alpha) variable. For logical expressions, you can use symbols or mnemonics as show in the table below.

Logical Comparison	Mnemonic	Symbol
Equal to	EQ	=
Not equal to	NE	^=
Less than	LT	<
Greater than	GT	>
Less than or equal to	LE	<=
Greater than or equal to	GE	>=

Once you click APPLY, the filter expression shows up next to the Filter icon. (See Figure 6.17.)

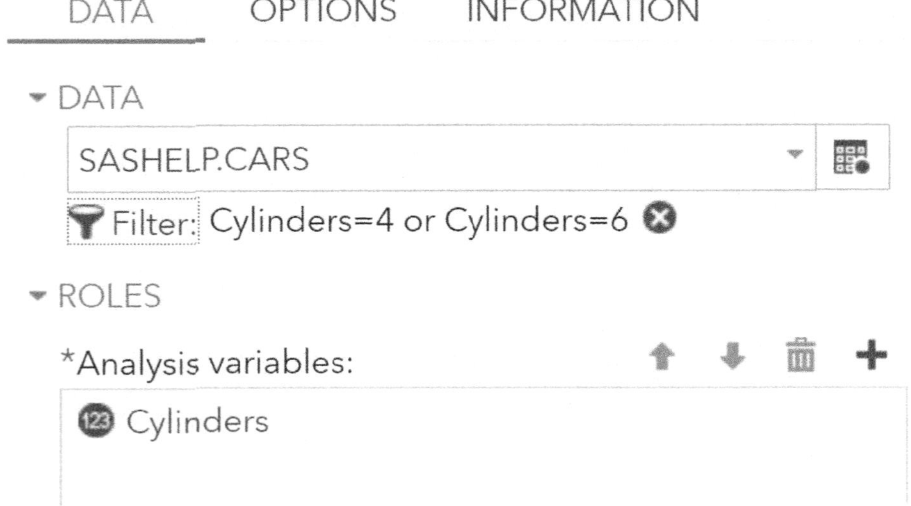

Figure 6.17: The Filter Condition Is Displayed

Now that you know how to set up a filter, let's continue with the Data Distribution task under the Statistics tab. To see a distribution of Horsepower for four- versus six-cylinder cars, select the variable Cylinders in the Group analysis box under the Additional Roles drop-down list in the DATA tab, as shown next.

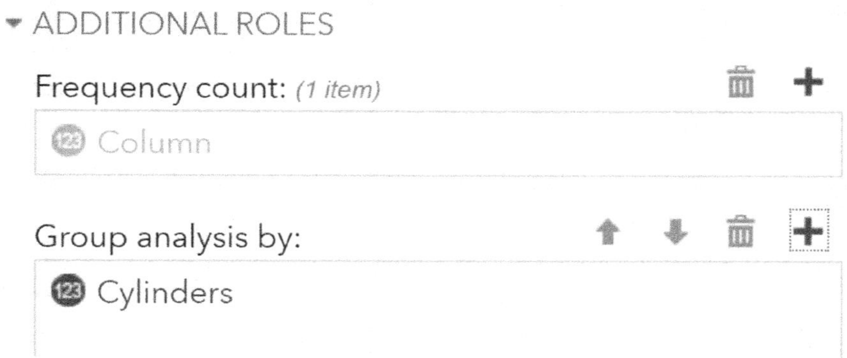

ADDITIONAL ROLES

Frequency count: *(1 item)*

Column

Group analysis by:

Cylinders

Figure 6.18: Choosing Your Grouping Variable

Make sure that you have checked the box next to Histogram on the OPTIONS tab and run the program. This produces a histogram of Horsepower for four- and six-cylinder cars. The histogram for four-cylinder cars is shown in Figure 6.19.

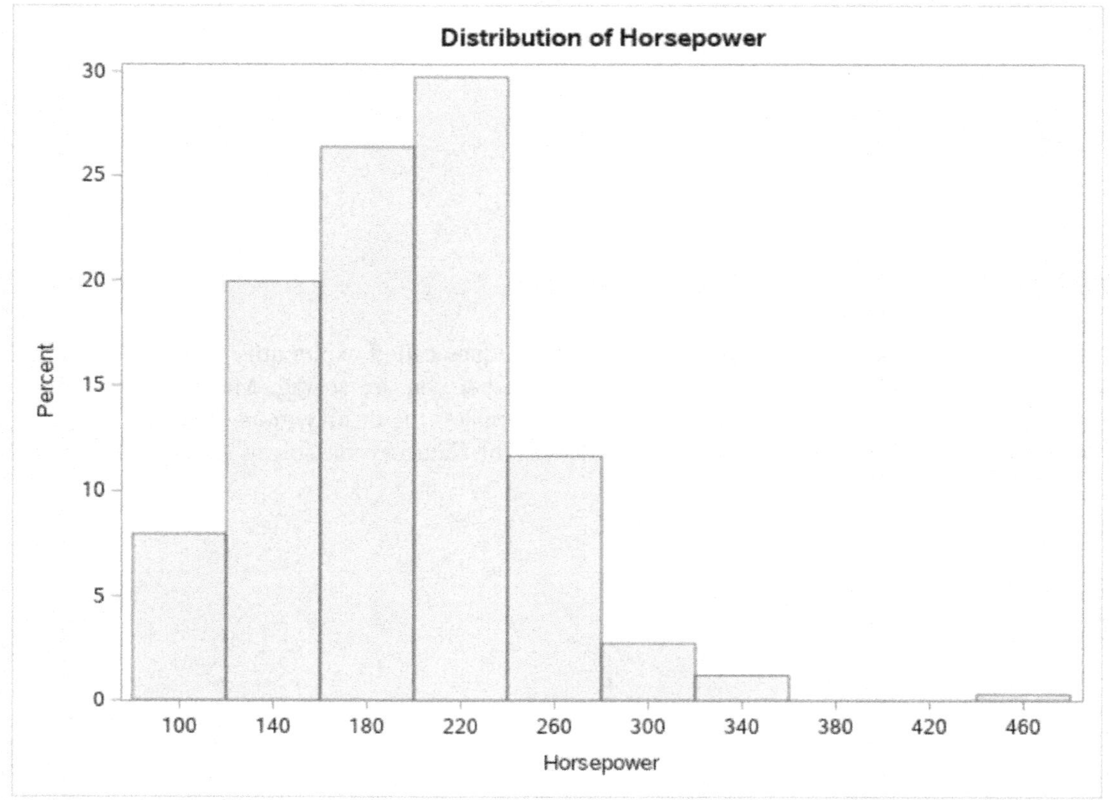

Figure 6.19: Distribution of Horsepower for Cars with Four Cylinders

Creating a Box Plot

An interesting way to display a single distribution or to see several distributions side-by-side is to use the Box Plot option on the Graph Task menu, as shown in Figure 6.20.

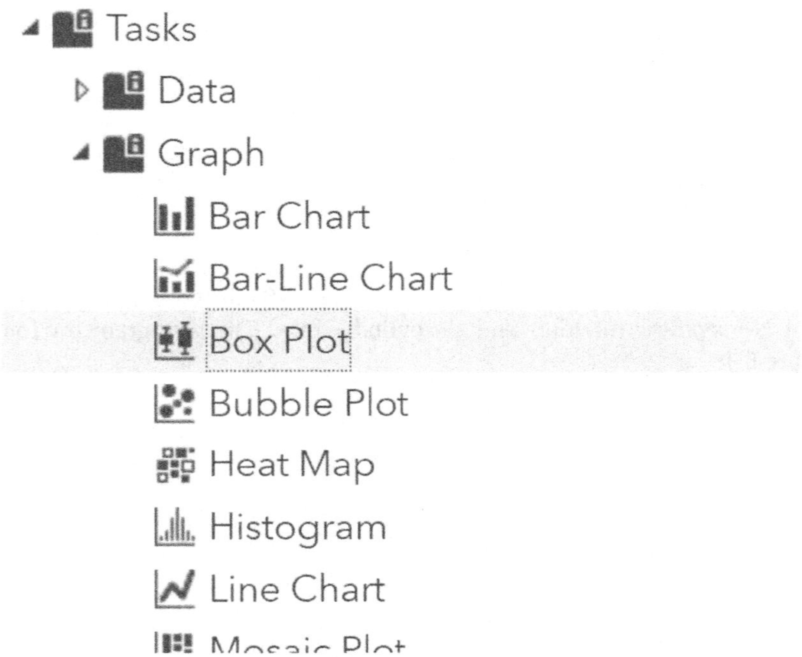

Tasks
Data
Graph
Bar Chart
Bar-Line Chart
Box Plot
Bubble Plot
Heat Map
Histogram
Line Chart

Figure 6.20: Selecting a Box Plot on the Graph Task

A box plot is a part of a collection of plots and techniques called exploratory data analysis (EDA). Let's run the Box Plot task first and then describe what you are seeing. Make sure you re-enter your filter criteria, select a horizontal orientation (this makes the display more compact), and choose Horsepower as the Analysis variable and Cylinders as the Category variable as shown in Figure 6.21.

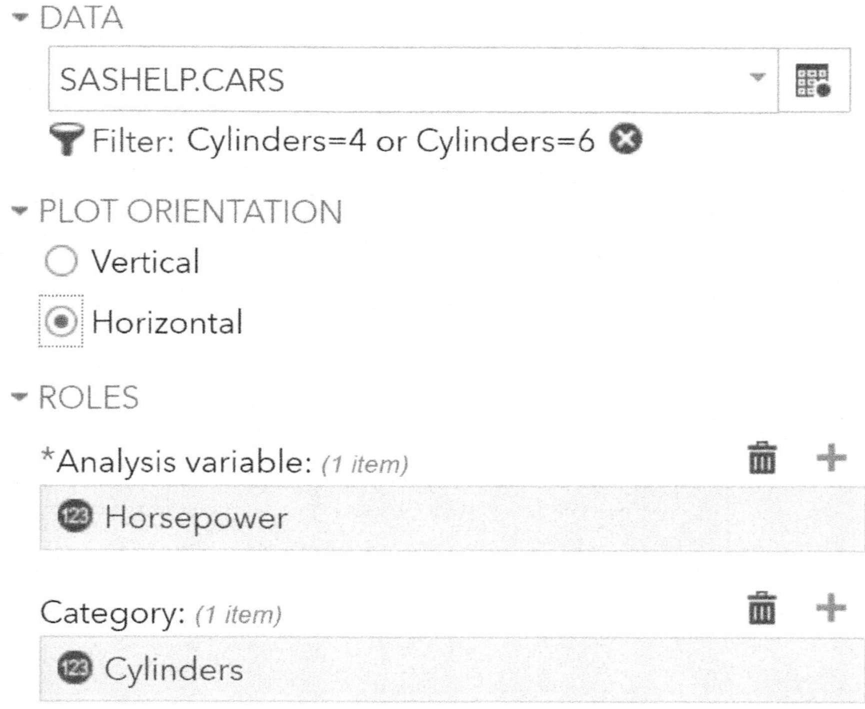

Figure 6.21: Requesting a Horizontal Box Plot for Horsepower Broken Down by Cylinders

It's time to run the task. The result is shown in Figure 6.22.

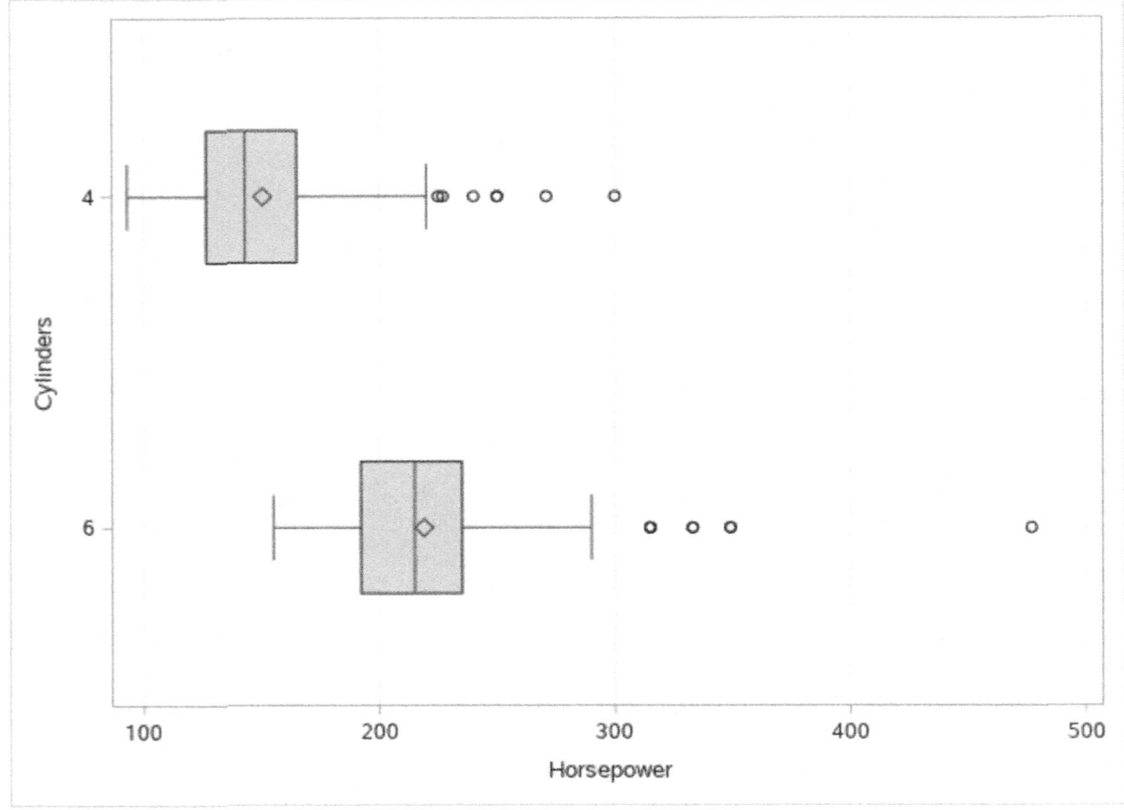

Figure 6.22: Box Plots Showing Horsepower for Four- and Six-Cylinder Cars

For each value of Cylinder, you see a box, lines coming out of each side of the box, and some small circles on the right side of the display. The vertical line within each box represents the median. Remember that half the values of your data fall below the median, and half the values in your data fall above the median. The median is also referred to as the 50th percentile. The small diamond in the box represents the mean. True believers of EDA would not include the mean; however, it is useful to see where the mean lies with respect to the median. Because both of these distributions are positively skewed (the tail is to the right), you expect the mean to be higher than the median, and that is confirmed by the box plots shown here. The left and right side of the box represents the 25th percentile and the 75th percentile, respectively. The 25th and 75th percentile are also referred to as the **first quartile** (Q1) and the **third quartile** (Q3). Notice that the box contains 50% of all the data values and the distance between Q1 and Q3 is called the **interquartile range**. Wow, that's a lot of terminology! The horizontal lines on the left and right side of the box represents any data values within 1.5 interquartile ranges below Q1 or above Q3. Finally, the small circles that you see on the display represent outliers—values that are more than 1.5 interquartile ranges above Q3 or below Q1. They represent the data that you saw in the right tail of the histograms shown earlier.

Conclusion

Before you conduct statistical tests on your data, it is a good idea to explore your data with the descriptive techniques (both tables and graphical output) described in this chapter. Knowing the shapes of distributions for continuous variables may affect your choice of statistical tests to perform. Frequency analysis will enable you to determine how many items (observations) belong to each category of a categorical variable. Both of these tasks also have the ability to uncover data errors.

Chapter 6 Exercises

1. Using the data set IRIS in the SASHELP library, generate summary statistics for variables PetalLength and PetalWidth. Include the mean, standard deviation, median, the number of nonmissing observations, and the number of missing observations. Using the tab ADDITIONAL ROLES, request the 95% confidence limits. Finally, request a histogram for the variable PetalLength.
2. Using the same data set in exercise 1, check for normality for the variable PetalLength. Add a Normal Quantile-Quantile (Q-Q) plot and compute skewness and kurtosis for this variable.
3. Using the same data set in exercise 1, compute one-way frequencies for the variable Species.
4. Using the same data set in exercise 1, generate a horizontal box plot for the variable PetalLength using Species as the category variable. Use a filter to remove the species "Virginica."
5. Using the data set Heart in the SASHELP library, investigate the distribution of the variable Weight, broken down by the variable Sex. Using the Anderson-Darling test for normality, would you reject the null hypothesis (at the .05 level) that Weight is normally distributed?
6. Using the data set Heart in the SASHELP library, produce a histogram and box plot for Weight and Height separately for males and females (variable Sex with values of F and M).
7. Using the data set Cars in the SASHELP library, create a horizontal box plot for Invoice (invoice price) for each value of the variable Cylinders. Do this again with a filter to restrict the rows in the table to 4 or 6 cylinders. Hint: Your filter expression should be: Cylinders = 4 or Cylinders=6.

7.

One-Sample Tests

Introduction

SAS Studio comes equipped with statistical tasks for just about any statistical query that you will need as a student or researcher.

You might have very little need to perform a one-sample test, but let's start there anyway. Typically, a **one-sample test** is used to determine if a single sample comes from a population with a known mean. This is in contrast to a two-sample test (discussed in the next chapter), where you want to test if two samples come from populations with different means. This chapter will show you how to navigate the various tabs that are common to all the statistical tasks. You will also see how to test some basic assumptions that need to be met before performing most parametric tests.

Getting an Intuitive Feel for a One-Sample t Test

Before we get into the details of running a one-sample *t* test, let's get an intuitive feel for how this test works.

The purpose of a one-sample test is usually to demonstrate that your sample of values comes or does not come from a population whose mean is known. For example, you might have before and after scores for students in a math program. You want to see whether they improved as a result of the program. If the null hypothesis was that the math program had no effect on scores, the mean difference (after – before) of the population from which you took your sample would be 0. Suppose you sample 10 students and the difference scores are as follows:

```
10 11 9 12 12 10 8 9 10 11
```

The mean is 10.2. Do you think the math program improved scores? You need to use these scores to estimate a population mean (your best guess would be 10.2, the mean of the sample). You need to determine whether it is unlikely to get a mean of 10.2 from a sample of 10 scores if the true population mean was 0. Your intuition should be that, given these 10 scores, that situation is very unlikely.

A *t* test is one way to assign a probability that, if the null hypothesis were true, you would obtain a sample mean as large or as small as you got, by chance alone. This is the famous *p*-value that you see reported in almost any study. If the probability is small (usually defined as less than .05), you reject the null hypothesis and accept the alternative hypothesis.

For those curious folks, running a one-sample *t* test on the 10 difference scores above results in a *p*-value of less than .0001. The math program worked.

Performing a One-Sample t Test

For this first example of a one-sample *t* test, we are going to visit a place called Small Town USA. This town has a single school for all grades—kindergarten through high school. Fifteen students in this school took the SAT exam in hopes of going to a quality college. The national average on the combined mathematics and English portions of the SAT is 1068. The 15 scores of the students who took the SAT from the Small Town School were entered into an Excel spreadsheet, as shown in Figure 7.1.

	A	B
1	Student	SAT_Score
2	1	1068
3	2	1558
4	3	1200
5	4	1402
6	5	1377
7	6	1035
8	7	1100
9	8	1111
10	9	1545
11	10	1323
12	11	1221
13	12	1000
14	13	1399
15	14	1232
16	15	1488
17		

Figure 7.1: SAT Scores from Small Town School (SAT_Scores.xlsx)

The school administrators want to know if their students scored higher than the national average. Even though the mean score for these 15 students was 1270.6, is it possible these students guessed really well on the test? If they were really similar to the rest of the country, what would be the chance of getting such a high mean?

To answer that question, you decide to perform a one-sample *t* test. The first step is to use the Import Data facility under the Tasks and Utilities tab as described in the previous chapter. Instead of clicking on the Select File box in the Import task, this time, for variety, you click the Server Files and Folders tab, then myfolder, click the SAT_Scores.xlsx file, and drag it over to the drop and drag area.

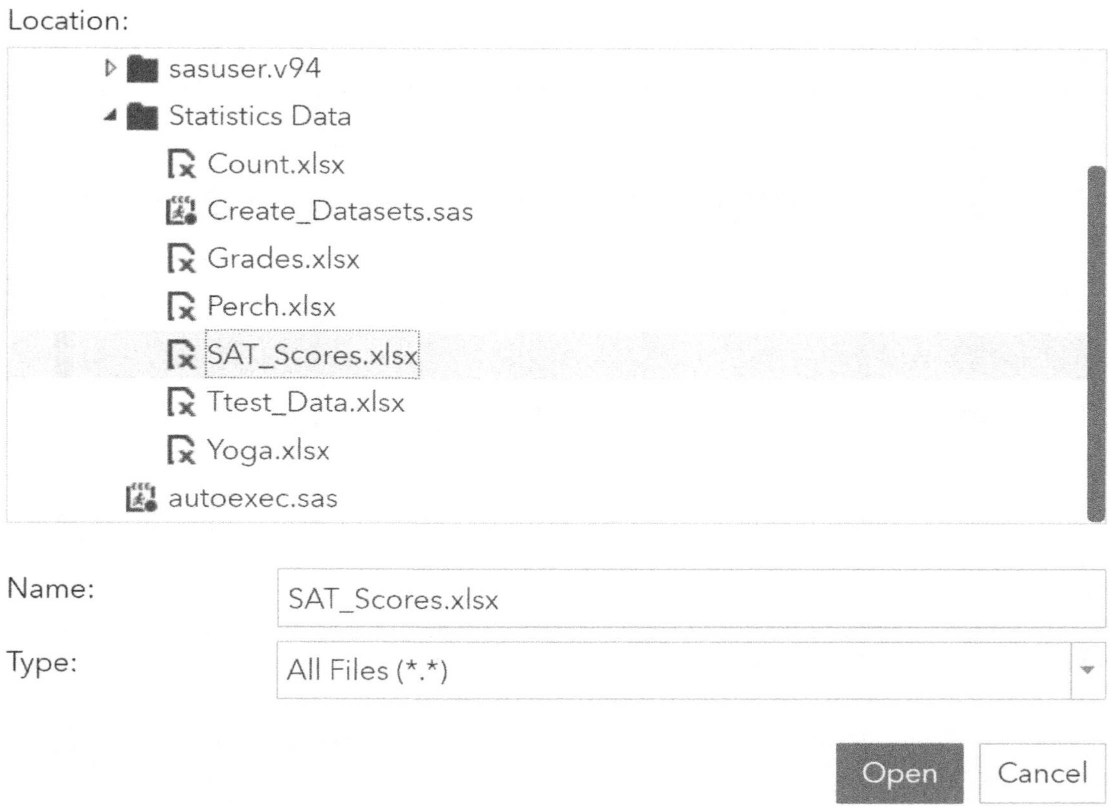

Location:
sasuser.v94
Statistics Data
Count.xlsx
Create_Datasets.sas
Grades.xlsx
Perch.xlsx
SAT_Scores.xlsx
Ttest_Data.xlsx
Yoga.xlsx
autoexec.sas
Name: SAT_Scores.xlsx
Type: All Files (*.*)
Open Cancel

Figure 7.2: Selecting your File in the Server Files and Folders Tab

This brings up the following screen:

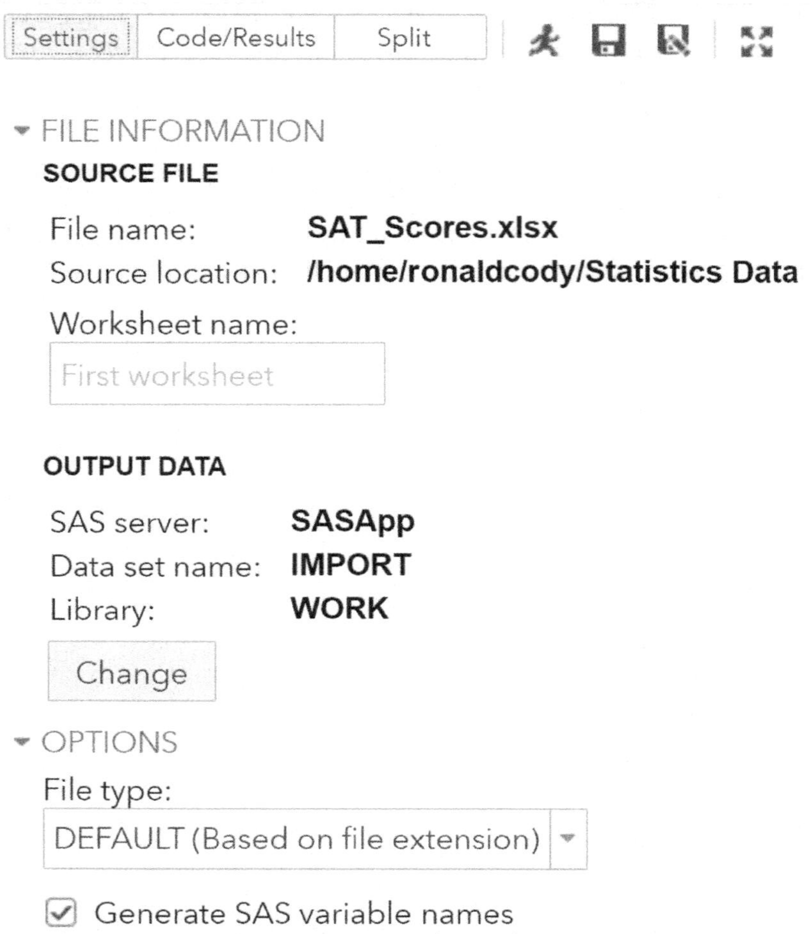

Figure 7.3: Importing the SAT_Scores Workbook

Click Change, name the SAS data set SAT_Scores, and place it in the WORK library.

> Remember, all files that you place in the WORK library disappear when you close your SAS session.

You are now ready to run your one-sample *t* test. Start by expanding Tasks on the navigation pane and then expand the Statistics tab. (See Figure 7.4.)

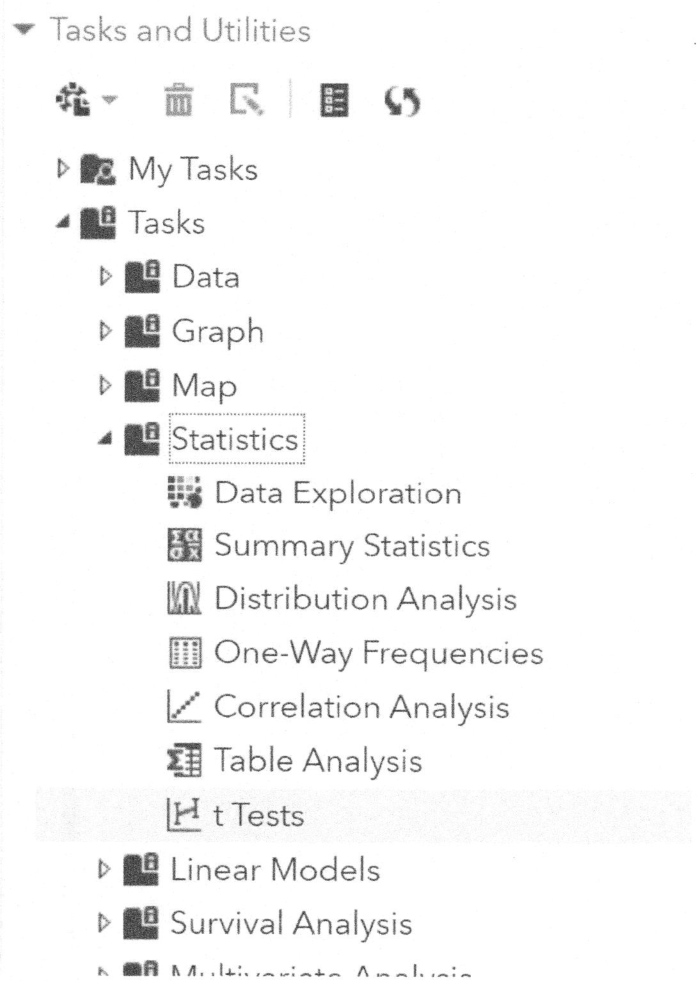

Tasks and Utilities

My Tasks
Tasks
Data
Graph
Map
Statistics
Data Exploration
Summary Statistics
Distribution Analysis
One-Way Frequencies
Correlation Analysis
Table Analysis
t Tests
Linear Models
Survival Analysis

Figure 7.4: Selecting a One-Sample Test

As with almost every statistical task, the first tab you see is the DATA tab. This tab is where you select the SAS data set that contains your data and select which variables you need for various roles. Figure 7.5 shows the DATA tab for the One-Way task.

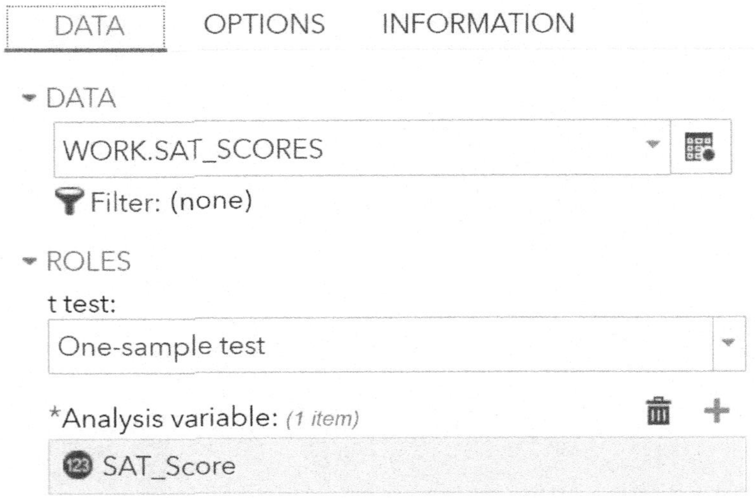

DATA OPTIONS INFORMATION

DATA

WORK.SAT_SCORES

Filter: (none)

ROLES

t test:

One-sample test

*Analysis variable: (1 item)

SAT_Score

Figure 7.5: The DATA Tab for One-Sample Tests

Select the data set SAT_Scores in the WORK library. In the Roles pull-down list, select a One-sample test and the variable SAT_Score as the Analysis variable. The next step is to open the OPTIONS tab (Figure 7.6).

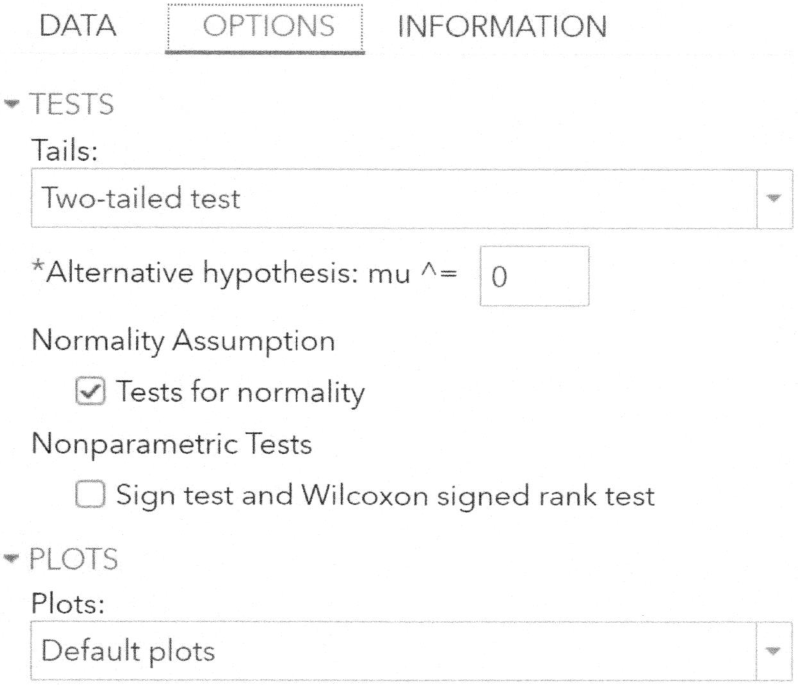

DATA OPTIONS INFORMATION

▾ TESTS

Tails:

Two-tailed test

*Alternative hypothesis: mu ^= 0

Normality Assumption

☑ Tests for normality

Nonparametric Tests

☐ Sign test and Wilcoxon signed rank test

▾ PLOTS

Plots:

Default plots

Figure 7.6: The OPTIONS Tab for One-Sample Tests

Because just about every test that you perform is a two-tailed (non-directional) test, you leave that default selection as it is. A two-tailed test can result in a significant result if the sample mean is significantly higher or lower than a hypothesized value. Even though you expect that these students have higher scores than the national average, you still allow for a result that they performed below the average. Next, you specify an Alternative hypothesis that the mean of the population from which you took your sample is not 1068, the national average. You might be more familiar with stating a null hypothesis instead of an alternative hypothesis when conducting one-sample *t* tests. However, specifying a null hypothesis as *mu* (the population mean) is equal to 1068 is equivalent to specifying the alternative hypothesis that *mu* is not equal to 1068. You should also check the box labeled Tests of normality because normality is one of the assumptions for performing a one-sample *t* test.

You can select which plots you would like the task to display, but for now, leave the selection of Default plots. It is time to run the task, so click the Run icon.

The first table displayed shows the results of four different tests for normality (Figure 7.7).

Variable: SAT_Score (SAT_Score)

Tests for Normality				
Test	**Statistic**		**p Value**	
Shapiro-Wilk	W	0.941123	Pr < W	0.3967
Kolmogorov-Smirnov	D	0.137436	Pr > D	>0.1500
Cramer-von Mises	W-Sq	0.045519	Pr > W-Sq	>0.2500
Anderson-Darling	A-Sq	0.305961	Pr > A-Sq	>0.2500

Figure 7.7: Several Tests for Normality

You will see slightly different results from these tests. Some of these tests are more or less likely to reject the null hypothesis that your data came from a population that was normally distributed. In most cases, all (or a majority) of the tests will lead you to the same conclusion. In this example, none of the four tests are significant at the .05 level (the magic number in statistics).

> Looking at the *p*-values from the tests of normality is not sufficient to make a determination that it is OK or not OK to conduct a *t* test.

Why is this important? (I will repeat this discussion in relation to several other statistical tests later in this book because it is so important.) If you have a very large sample size, all of these tests for normality might be significant, even though your distribution is close to a normal distribution. This is because statistical tests with large sample sizes have more power to detect small differences. If your sample size is small, the tests for normality can often fail to reject the null hypothesis, and it is with small samples that the normality assumption is important.

Here is the problem: Most of the statistical tests that we discuss in this book have as one of their assumptions that you have normally distributed data. However, the central limit theorem states that the sampling distribution will be normally distributed if n, the sample size, is sufficiently large. Your next question should be: What is sufficiently large?

A sample size that is considered sufficiently large depends on the shape of the distribution of values. If the distribution is somewhat symmetrical, sufficiently large might be quite small (10 or 20). If the distribution is highly skewed, sufficiently large might be quite large. Before you decide to abandon the one-sample *t* test, you should look at the distribution of your scores in a histogram and/or a Q-Q plot. The one-sample *t* test task produces both of these plots to help you understand how your data values are distributed.

An aside: When I worked as a biostatistician at a medical school in New Jersey, I would consult with researchers, many of whom had some statistical expertise. They would show me output from a test (such as a *t* test) and point out that the test of normality rejected the null hypothesis and that they could not use a *t* test (or other parametric test) to analyze their data. My next question would be, "What is your sample size?" Sometimes the answer would be hundreds or thousands. If that is the case, and the distribution as shown by a histogram is mostly symmetrical, you can use the results of the *t* test (or other test requiring normally distributed data) without any problem.

The next part of the output shows the mean and standard deviation along with several other statistics. Of particular interest are the *t* value and the probability of a type I error.

Variable: SAT_Score (SAT_Score)

N	Mean	Std Dev	Std Err	Minimum	Maximum
15	1270.6	186.4	48.1202	1000.0	1558.0

Mean	95% CL Mean		Std Dev	95% CL Std Dev	
1270.6	1167.4	1373.8	186.4	136.4	293.9

DF	t Value	Pr > |t|
14	26.40	<.0001

Figure 7.8: t Test Results

You see that the mean SAT score for your 15 students is 1270.6, and the *p*-value is .0009 (often described as highly significant).

Even though none of the tests for normality were significant, you still want to inspect the histogram (Figure 7.9) and Q-Q plot (Figure 7.10).

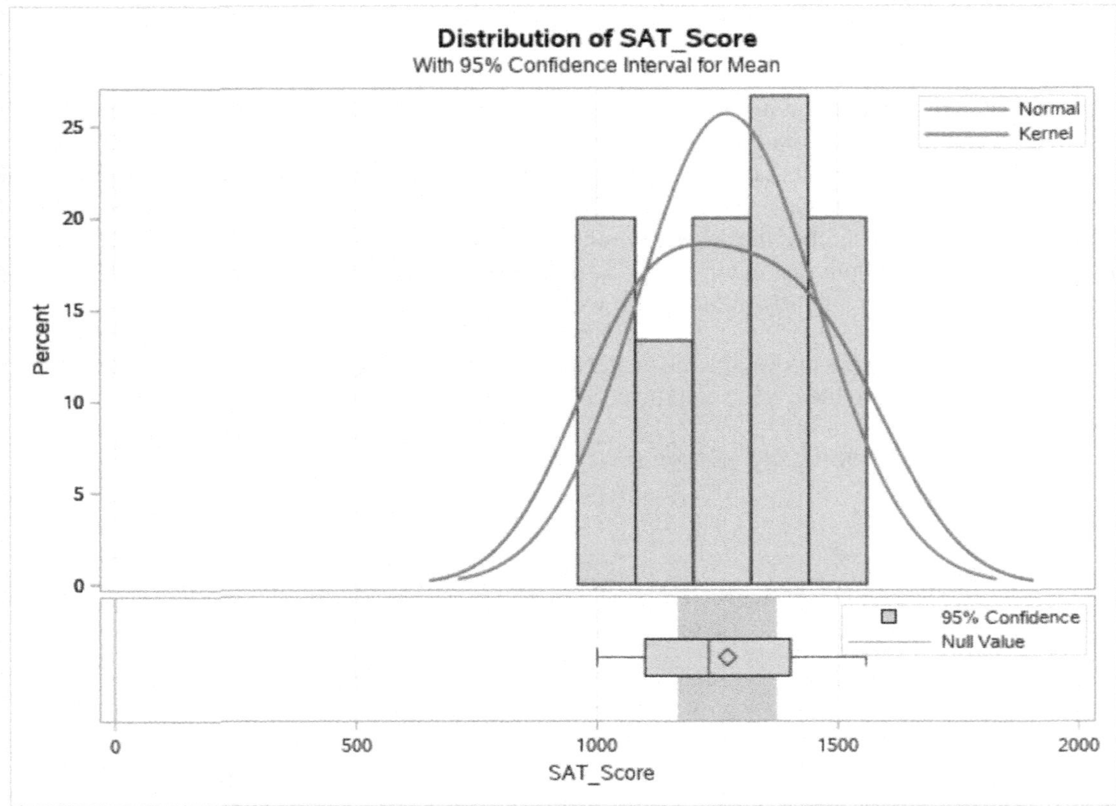

Figure 7.9: Histogram and Box Plot

Even though this doesn't resemble a normal distribution, it is symmetric and, with a sample size of 15, you feel confident that a *t* test was appropriate.

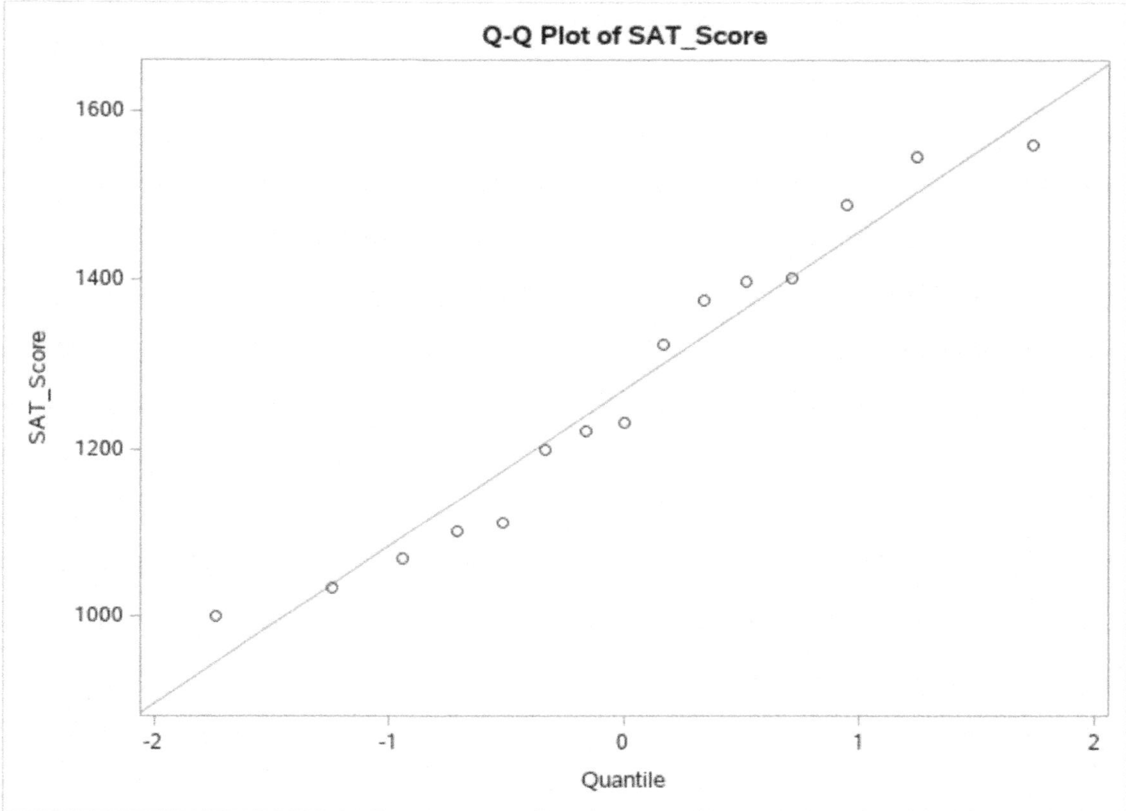

Figure 7.10: Q-Q Plot

As you (hopefully) remember from Chapter 6, data values from a normal distribution would lie along the diagonal line. This plot does not show any strong deviations from a straight line and confirms the decision that you made by inspecting the histogram.

Nonparametric One-Sample Tests

If you do not feel that a *t* test is appropriate because of a non-normal distribution (especially if your sample size is relatively small), there is a nonparametric alternative called the Wilcoxon Signed Rank Test. To demonstrate how this test works, we will use a data set called Before_After that is included in the Create_Datasets.sas program. Remember, if you closed your SAS session, regardless of what platform you are using, you need to re-run the program to create all the WORK data sets.

The data set Before_After contains difference scores (variable name Difference) on a performance test, before and after several training sessions. Go ahead and run the *t* Tests task and inspect the tests of normality.

Variable: Difference

Tests for Normality				
Test	**Statistic**		**p Value**	
Shapiro-Wilk	W	0.769706	Pr < W	0.0015
Kolmogorov-Smirnov	D	0.304244	Pr > D	<0.0100
Cramer-von Mises	W-Sq	0.280147	Pr > W-Sq	<0.0050
Anderson-Darling	A-Sq	1.471368	Pr > A-Sq	<0.0050

Figure 7.11: Tests of Normality for the Variable Difference

Because there are only 15 observations in this data set and all the tests for normality are significant, you proceed to look at the histogram.

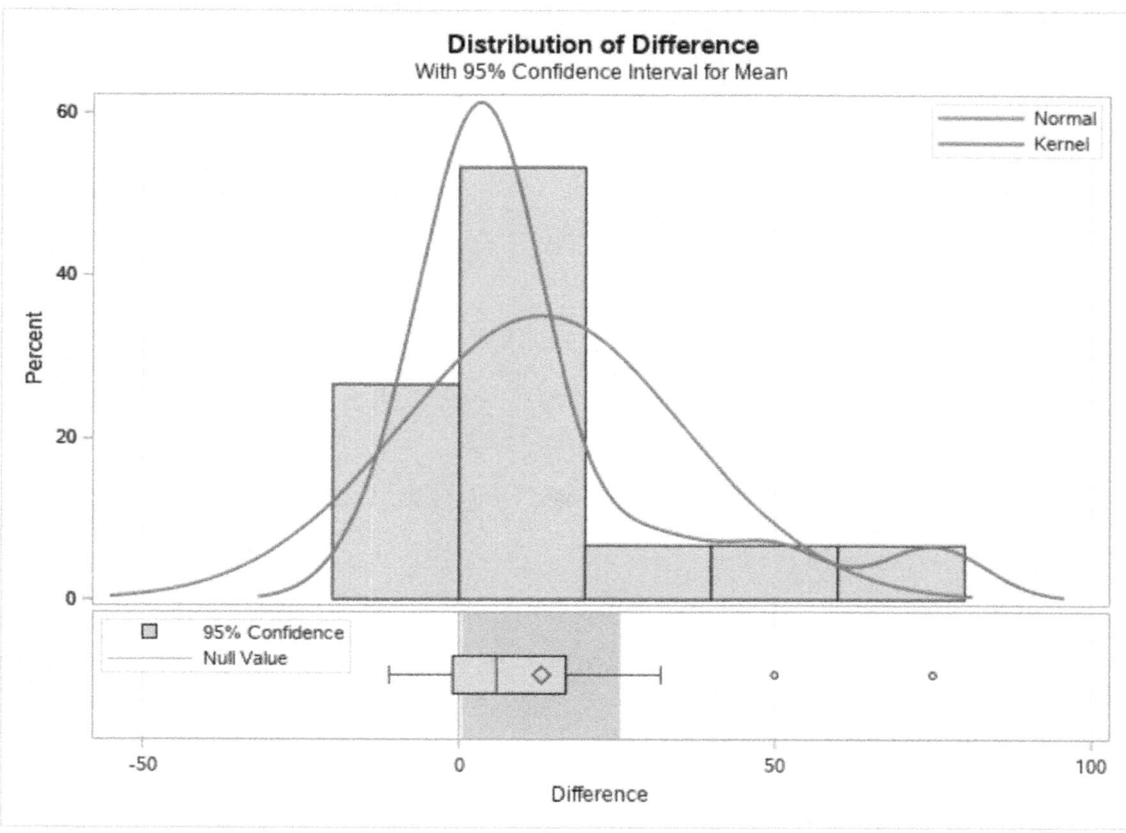

Figure 7.12: Histogram and Box Plot

You decide that a *t* test is not appropriate and decide to run a nonparametric test. A nonparametric test does not require the assumption of normally distributed data. To run a Wilcoxon Signed Rank test, go to the OPTIONS tab and check the box under Nonparametric tests for Sign test and Wilcoxon signed rank test, as shown in Figure 7.13.

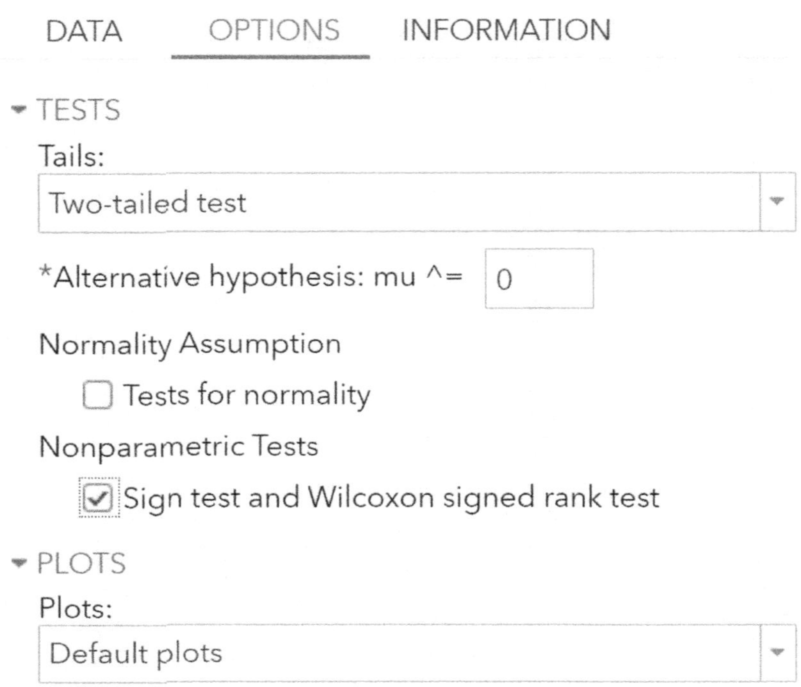

DATA OPTIONS INFORMATION

TESTS

Tails:

Two-tailed test

*Alternative hypothesis: mu ^= 0

Normality Assumption

☐ Tests for normality

Nonparametric Tests

☑ Sign test and Wilcoxon signed rank test

PLOTS

Plots:

Default plots

Figure 7.13: OPTIONS Tab for One-Sample t Tests

It's time to run the procedure. Even though you selected an option for nonparametric tests, the task also runs a traditional *t* test as well. You can ignore that portion of output. For the curious, the *p*-value from the *t* test was .0431. Even though it is less than the traditional value of .05 for significance, you should ignore it. You want to scroll to the bottom of the results to see the *p*-value from the two nonparametric tests of interest, the Sign Test and the Wilcoxon Signed Rank Test.

Variable: Difference

Tests for Location: Mu0=0				
Test	Statistic		p Value	
Student's t	t	2.223707	Pr > |t|	0.0431
Sign	M	3	Pr >= |M|	0.1796
Signed Rank	S	36.5	Pr >= |S|	0.0183

Figure 7.14: Nonparametric Tests on the Difference Score

Of the two tests, the Signed Rank Test is usually the one you want to report. What follows is a description of both of these nonparametric tests.

The **Sign Test** simply looks to see whether the difference score is positive or negative. Under the null hypothesis, there should be an equal number of positive and negative values. The *p*-value for this test is based on a binomial probability. The *p*-value listed here is .1796 (that is, not significant). This test is most often used when the scores or differences can only be judged to be positive or negative, and a numerical value cannot be placed on the difference.

By comparison, the **Signed Rank Test** usually has more power to detect differences and is preferred if you have a measure of the differences. To conduct a Signed Rank Test, take the absolute value of all your differences (that is, ignore any minus signs) and rank them. If the difference is zero, ignore that value. Next determine the sign for each rank, based on the original values. If the null hypothesis is true, the positive and negative ranks should be about equal. If most of the higher ranks are of the same sign, you might have evidence to reject the null hypothesis. The probability is listed as .0183 (significant at the .05 level).

Conclusion

One of the advantages of running SAS Studio is the ease with which you can perform a large number of statistical tests. Yes, you still need to understand which tests to run and verify that the assumptions for those tests are satisfied. But once you have done this, getting your results is a few mouse clicks away.

Chapter 7 Exercises

1. Using the SAS data set Heart in the SASHELP library, use a one-way *t* test to determine if the mean weight of the population from which the sample was drawn is equal to 150 pounds. Include a test of normality and generate a histogram and box plot. Should you be concerned that the tests of normality reject the null hypothesis at the .05 level?
2. Using the data set Fish in the SASHELP library, test if the mean weight of Smelt (use a filter: Species = 'Smelt')) is equal to 10. Be sure to run both a parametric and nonparametric test for this analysis. How do the parametric and nonparametric tests compare? Would you reach the same conclusion using these tests?
3. Using the SAS data set Air in the SASHELP library, do the following:
   a. List the first 10 observations from this data set.
   b. Run summary statistics on the variable Air (number of flights per day in thousands). Include a histogram and box plot.
   c. Run a parametric and nonparametric one-way *t* test to determine if the mean number of flights per day (in thousands) is significantly different from 285.
4. Run the short program below and then run a one-way *t* test to determine if the difference scores (Diff) come from a population whose mean is zero. In this program, the RAND function is generating uniform random numbers (numbers between 0 and 1 (all with equal likelihood). The DO loop generates 20 of these random numbers and outputs them to the Difference data set.

> If you don't like to type, this program and all the other programs associated with the exercises are included in the program Create_Datasets.sas.

```
*Data set Difference;
data Difference;
   call streaminit(13579);
   do Subj = 1 to 20;
      Diff = .6 - rand('uniform');
      output;
   end;
run;
```

5. Rerun exercise 4, but change the value of 20 in the DO loop to 200. What do the tests for normality show? Is it OK to use a *t* test anyway? How did the *p*-value change?

8.

Two-Sample Tests

Introduction

There are two classes of two-sample tests that you will see in this chapter. One is an **unpaired *t* test** (also called a *t* test for independent groups) that is used to compare means between two groups. The other type of two-sample *t* test is a **paired *t* test**. Commonly, a paired *t* test is performed when each subject is measured twice, typically before and after some treatment. It can also be used when experimental units are paired on certain characteristics such as gender and age before the experiment, and the variable of interest is the difference between values in each pair. Experimental units are often subjects, and that is the terminology subsequently used in the book.

As you saw with the one-sample *t* test, the two-sample statistical task also performs nonparametric tests, both for paired and unpaired situations.

Getting an Intuitive Feel for a Two-Way t Test

Much of the same logic that we discussed in a one-sample *t* test applies here. Suppose you want to compare two treatments for back pain. One method is physical therapy, the other, bed rest. You collect 10 pain scores (on a scale of 1 to 10 with 1 being almost pain free and 10 being excruciating pain) for each treatment as follows:

```
Physical Therapy: 2 4 3 5 4 2 1 3 5 3 Bed Rest: 3 5 4 4 3 6 5 4 6 2
```

The mean for the physical therapy group is 3.2 and the mean for the bed rest group is 4.2. Do you think the difference is significant at the .05 level? The null hypothesis is that the physical therapy scores and the bed rest scores come from populations with equal means. How likely is it to get a difference of 1 point between two groups of 10 subjects if the null hypothesis were true? The answer, provided by a two-sample *t* test, is slightly greater than 10% (p = .1066). It might be the case that physical therapy is better than bed rest, but the probability that the difference occurred by chance is higher than the traditional 5%. Therefore, you will fail to reject the null hypothesis. Remember, this doesn't mean that the two treatments are equal—it just means that you didn't have enough confidence to reject a hypothesis of equality.

Unpaired t Test (t Test for Independent Groups)

The data for this example is located in a permanent SAS data set called Reading in the STATS library. If you have not already run the Create_Datasets.sas program, do it now by clicking Server Files and Folders. (See Figure 8.1.)

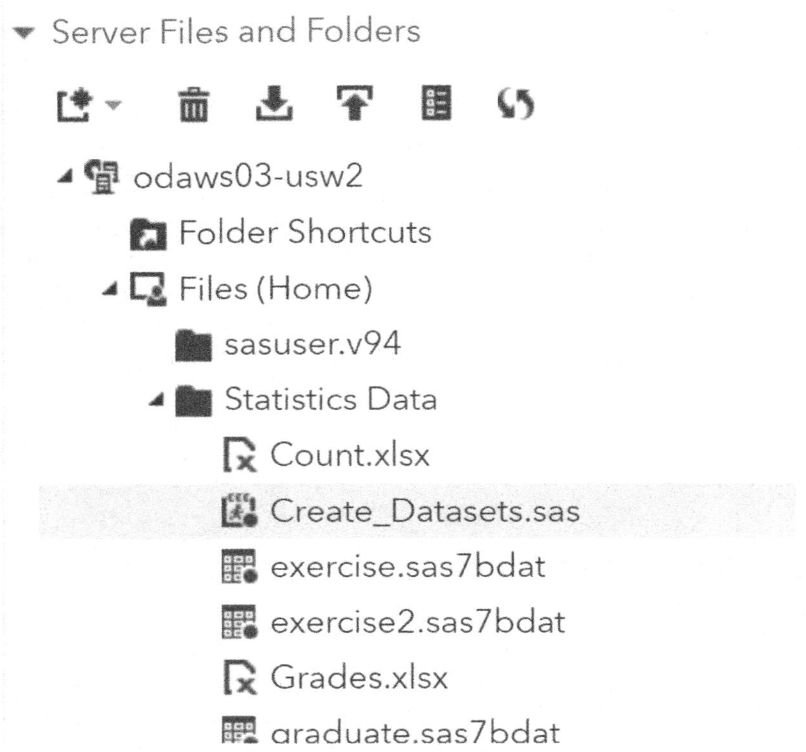

Figure 8.1: The Server Files and Folders Tab

If you double-click Create_Datasets.sas, you will see the program in the Code window. A portion of this window is shown in Figure 8.2.

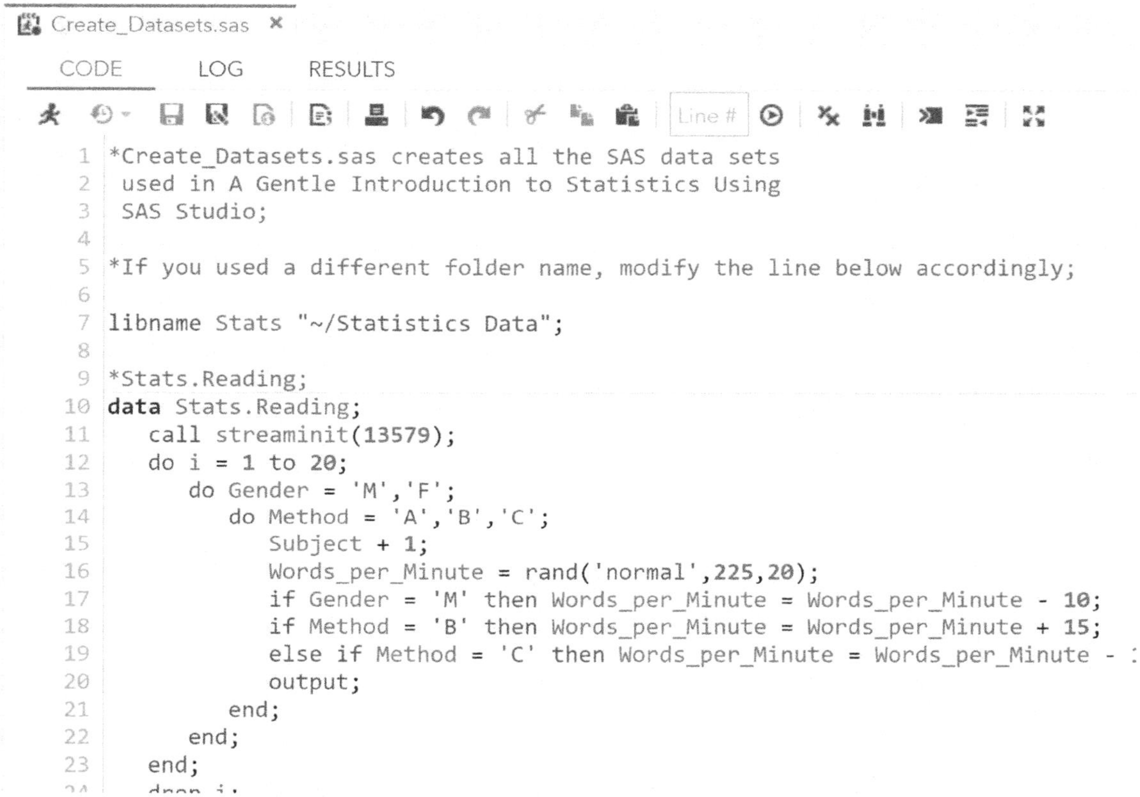

Create_Datasets.sas ×

CODE LOG RESULTS

Line #

```
*Create_Datasets.sas creates all the SAS data sets
 used in A Gentle Introduction to Statistics Using
 SAS Studio;

*If you used a different folder name, modify the line below accordingly;

libname Stats "~/Statistics Data";

*Stats.Reading;
data Stats.Reading;
   call streaminit(13579);
   do i = 1 to 20;
      do Gender = 'M','F';
         do Method = 'A','B','C';
            Subject + 1;
            Words_per_Minute = rand('normal',225,20);
            if Gender = 'M' then Words_per_Minute = Words_per_Minute - 10;
            if Method = 'B' then Words_per_Minute = Words_per_Minute + 15;
            else if Method = 'C' then Words_per_Minute = Words_per_Minute - :
            output;
         end;
      end;
   end;
   drop i;
```

Figure 8.2: Part of the CODE Window in SAS Studio

When you click the Run icon, all the data sets used in this book (other than the ones from the SASHELP library) will be created. This same program will also create all the data sets you will need for the end-of-chapter exercises.

Now that you have all the administrative tasks finished, let's discuss the data in the Reading data set. This data set contains fictional (made-up) data that shows reading speed (in words per minute), gender (M or F), and reading methods (A, B, or C).

Describing a Two-Sample t Test

Two-sample *t* tests are used to compare the means of two groups. Some text books refer to a *t* test by the longer name, Student's *t* Test. The history of the name is interesting. The test was devised in 1908 by William Gosset, who worked for the Guinness Brewery Company in Dublin, Ireland. It was the policy of the Guinness Company that none of its employees were allowed to publish any of the methods used by Guinness. Therefore, Gosset published his paper describing the *t* test under the pseudonym Student.

OK, enough of the history, I found that an interesting tidbit! There are a number of assumptions that need to be met before you perform this statistical test. One is that the scores in each group are normally distributed. As you saw in the one-sample *t* test, that assumption is a function of how much the distribution of scores deviates from a normal distribution and the sample size. Another assumption is that the variances (square of the standard deviation) are equal in the two groups. If that is not the case, there are options to adjust the *t* test.

Because there are only two genders in the data set, you can use a *t* test to compare the mean reading speeds for males and females. Let's get started:

The first step is to open the *t* Tests tab (the same process that you used for the one-sample *t* test):

Tasks and Utilities ⇒ Statistics ⇒ t Tests (double-click here)

This opens the following screen.

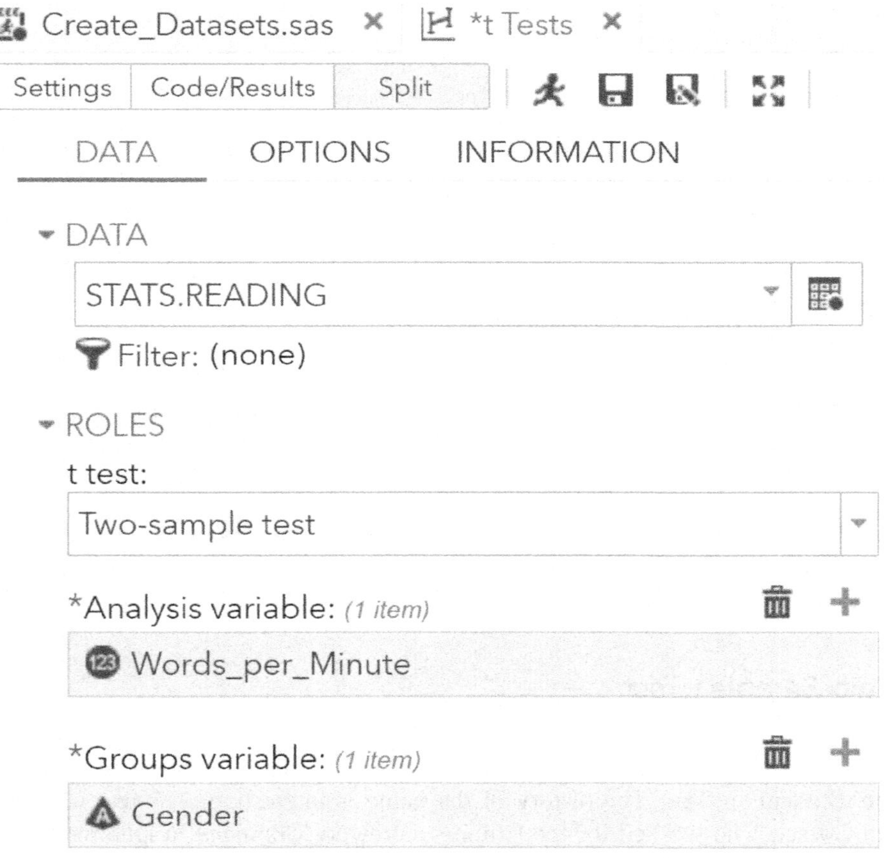

Figure 8.3: DATA Tab for Two-Sample t Test

Select the data set Reading from the STATS library. In the pull-down list, select Two-sample test. Select Words_per_Minute as the analysis variable and Gender as the Groups variable. (See Figure 8.3.)

The OPTIONS tab (Figure 8.4), enables you to select a one- or two-tailed test (you almost always want two-tailed) and select a test for normality.

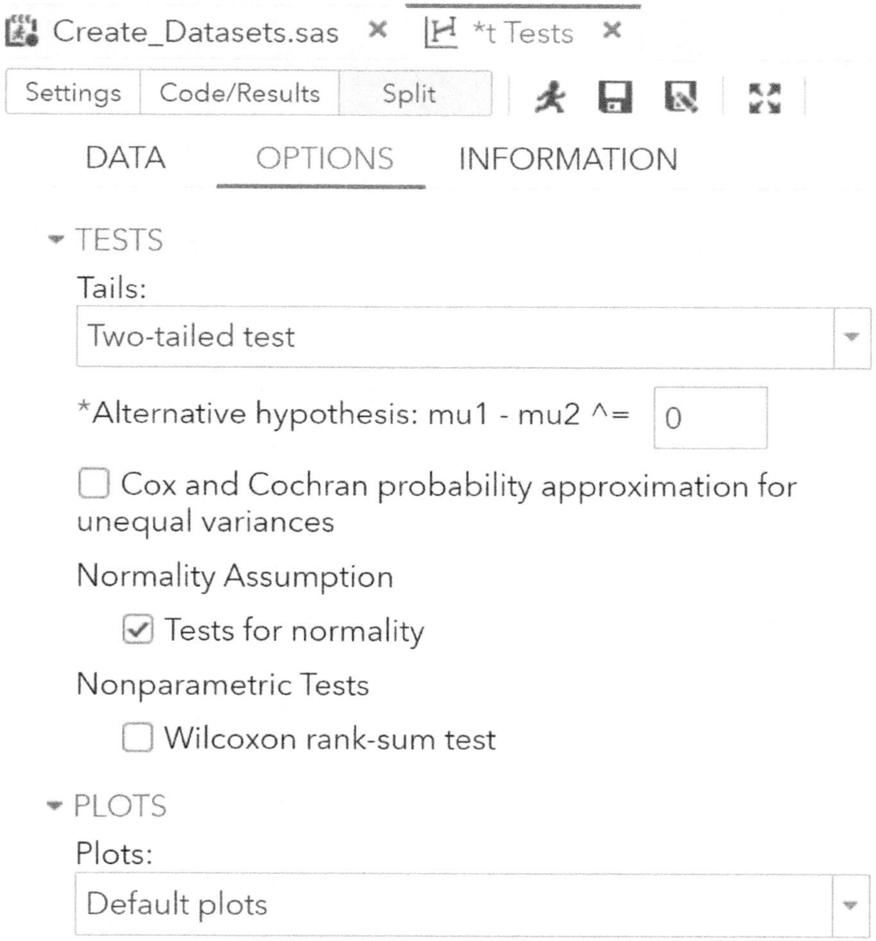

Create_Datasets.sas × *t Tests ×

Settings Code/Results Split

DATA OPTIONS INFORMATION

TESTS

Tails:

Two-tailed test

*Alternative hypothesis: mu1 - mu2 ^= 0

Cox and Cochran probability approximation for unequal variances

Normality Assumption

Tests for normality

Nonparametric Tests

Wilcoxon rank-sum test

PLOTS

Plots:

Default plots

Figure 8.4: The OPTIONS Tab for t Tests

At the bottom of this tab, you can select a variety of plots by using the menu under Plots. In this example, you are choosing Selected plots. It looks like this:

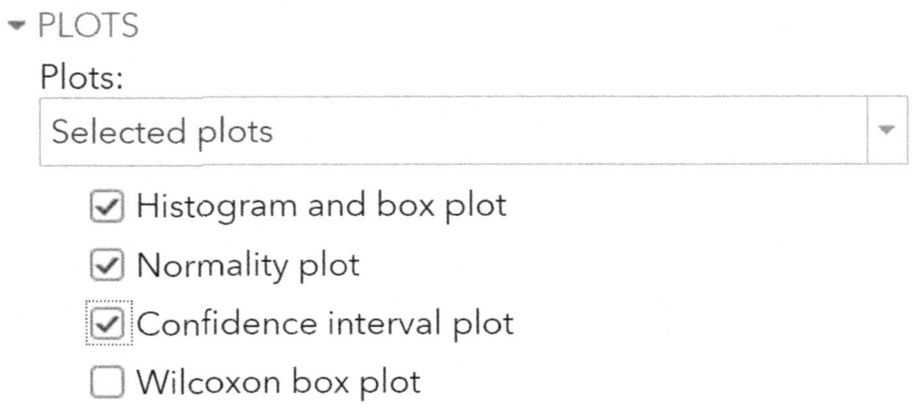

▾ PLOTS

Plots:

Selected plots

☑ Histogram and box plot

☑ Normality plot

☑ Confidence interval plot

☐ Wilcoxon box plot

Figure 8.5: Plot Options

You can select or deselect any of these plots. Here you are requesting the first three plots (and leaving off the Wilcoxon box plot). It's time to run the task. Here is the first part of the output:

Variable: Words_per_Minute
Gender = F

Tests for Normality				
Test	**Statistic**		**p Value**	
Shapiro-Wilk	W	0.989969	Pr < W	0.9044
Kolmogorov-Smirnov	D	0.051513	Pr > D	>0.1500
Cramer-von Mises	W-Sq	0.022901	Pr > W-Sq	>0.2500
Anderson-Darling	A-Sq	0.169494	Pr > A-Sq	>0.2500

Variable: Words_per_Minute
Gender = M

Tests for Normality				
Test	**Statistic**		**p Value**	
Shapiro-Wilk	W	0.992811	Pr < W	0.9786
Kolmogorov-Smirnov	D	0.058968	Pr > D	>0.1500
Cramer-von Mises	W-Sq	0.027792	Pr > W-Sq	>0.2500
Anderson-Darling	A-Sq	0.184146	Pr > A-Sq	>0.2500

Figure 8.6: Tests of Normality for Females and Males

This is similar to what you saw in the previous chapter when we performed a one-sample *t* test. Remember not to be too concerned with the *p*-values from the tests of normality. In this case, all the *p*-values are greater than the traditional cutoff of .05, meaning that you do not reject the null hypothesis that the data values come from a population that is normally distributed. It is more important to look at the histograms and pay attention to the sample size. Let's inspect the next part of the output:

Variable: Words_per_Minute

Gender	Method	N	Mean	Std Dev	Std Err	Minimum	Maximum
F		60	225.9	23.5454	3.0397	176.4	281.6
M		60	216.2	22.0345	2.8446	161.4	269.0
Diff (1-2)	Pooled		9.7073	22.8025	4.1631		
Diff (1-2)	Satterthwaite		9.7073		4.1631		

Gender	Method	Mean	95% CL Mean		Std Dev	95% CL Std Dev	
F		225.9	219.9	232.0	23.5454	19.9579	28.7175
M		216.2	210.5	221.9	22.0345	18.6772	26.8747
Diff (1-2)	Pooled	9.7073	1.4631	17.9514	22.8025	20.2274	26.1348
Diff (1-2)	Satterthwaite	9.7073	1.4627	17.9518			

Figure 8.7: Means, Standard Deviations, and Other Statistics

Here you see there are 60 Females and 60 Males. The mean reading speed for females and males is 225.9 and 216.2, respectively, with a mean difference of 9.7073. The 95% confidence limits for the difference in the means are 1.4631 to 17.9514. Because this interval does not include zero (a value indicating there is no difference in reading speed between men and women) the difference is significant at the .05 level.

The next table (Figure 8.8) shows the *t* values and probabilities for two assumptions. One probability, called Pooled, is for the assumption of equal variances. The other, called Satterthwaite, is a value computed assuming unequal variances. Which one should you choose?

Method	Variances	DF	t Value	Pr > |t|
Pooled	Equal	118	2.33	0.0214
Satterthwaite	Unequal	117.48	2.33	0.0214

Figure 8.8: T Table for Assumptions of Equal or Unequal Variances

Equality of Variances				
Method	Num DF	Den DF	F Value	Pr > F
Folded F	59	59	1.14	0.6121

Figure 8.9: Test for the Assumption of Equal Variances

There are different thoughts on how to approach this issue. The first, and probably the most common strategy, is to look at the *F* test in Figure 8.9. The *F* test is used to test the null hypothesis that the variances are equal. The alternative hypothesis is that they are not equal. If the *p*-value for this test is less than .05, you would choose the *t*- and *p*- values under the assumption of unequal variances (the method used is labeled Satterthwaite in the table). If the *p*-value is greater than .05, you would use the values based on equal variances (labeled Pooled in the table).

Another train of thought says to decide before you conduct your study which assumption (equal or unequal variances) you think is reasonable. The idea behind this method is that collecting the data and making the decision based on the same data is somewhat circular. The good news is that for large samples, the *t* test is very robust to the homogeneity of variance assumption, and the two *p*-values are usually close, even if there are relatively large differences in the variances.

Following rule one in this discussion, you see a *p*-value of 0.6121 for the test of homogeneity of variance and decide to use the pooled values in the table. Using a significance level of .05, you reject the null hypothesis that the populations from which you took the female and male reading speeds have equal means, and you conclude that the females read faster than males. Remember, the data for this section was simulated and not based on actual reading speeds.

The next portion of the output (Figure 8.10) shows histograms with an overlay of a normal and kernel distribution for females and males. A **kernel distribution** is a nonparametric method that attempts to fit a smooth curve using data points along the distribution. At this point, all you are concerned with is the general shape of the distribution and if it *looks* like it is highly skewed or not.

Underneath the histograms are two box plots. You learned about them in an earlier discussion in Chapter 6. Notice that there are two outliers, one to the left and one to the right on the box plot for males.

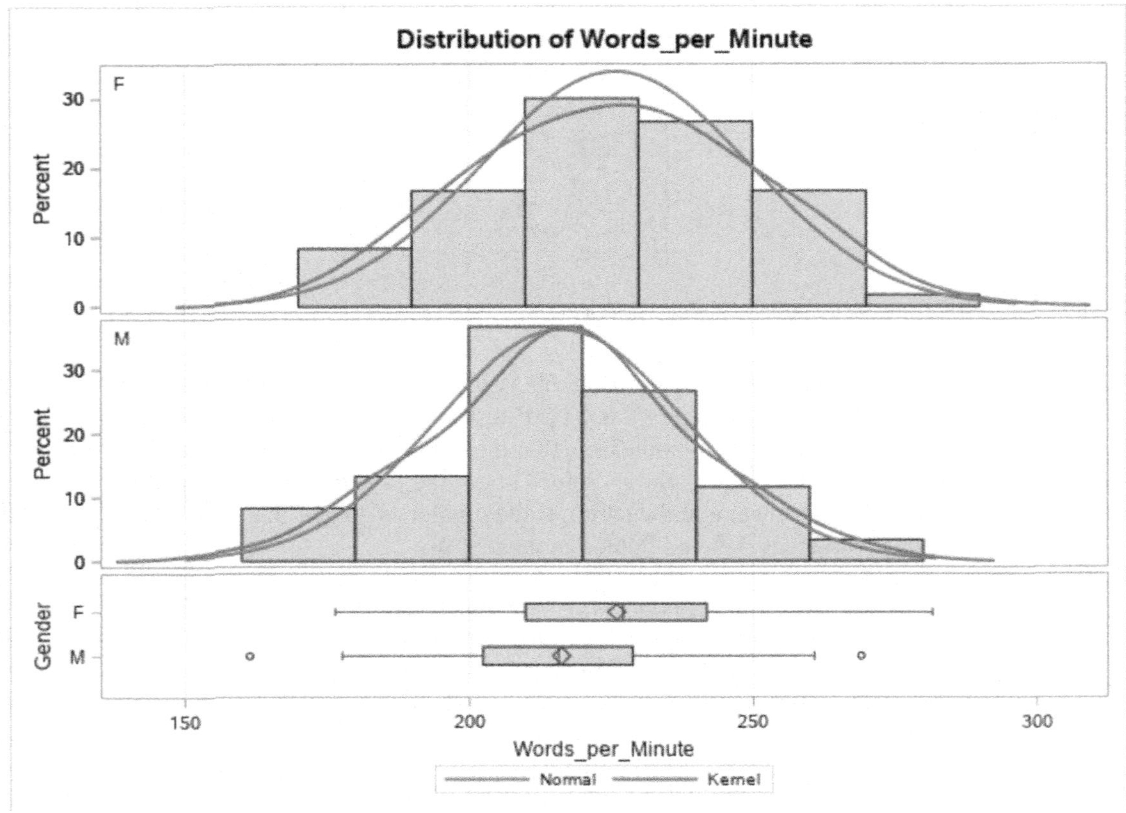

Figure 8.10: Histograms, Box Plots, and Normal Curves for Females and Males

The next portion of the output shows the mean difference between the groups along with the 95% confidence limits for the difference. You see one for the assumption of equal variance and one that is adjusted in case the variances are not equal. Notice that the mean differences and the confidence limits are almost identical.

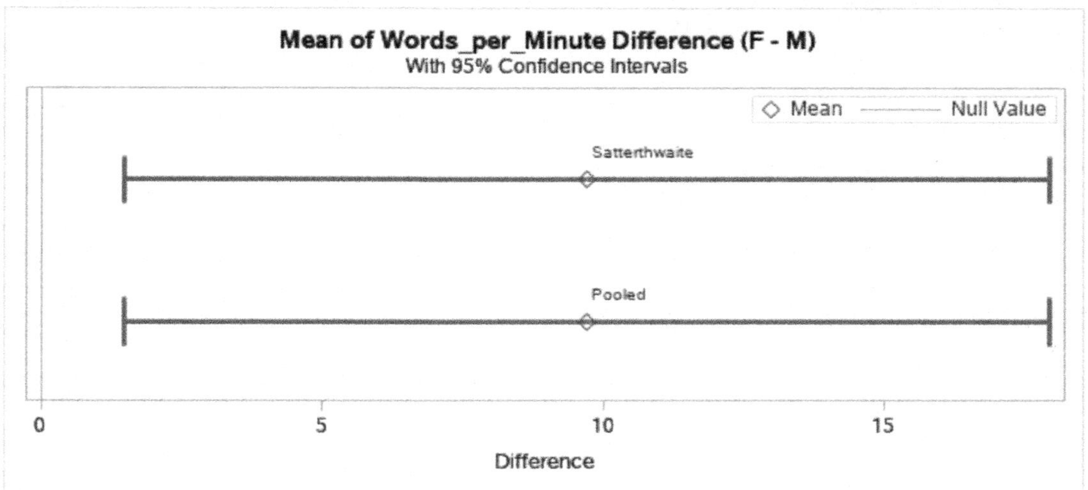

Figure 8.11: Mean Differences and the 95% Confidence Limits

The last portion of the output presents two Q-Q plots. Remember that you are looking for most of the data points to follow the straight line that represents data from a normal distribution.

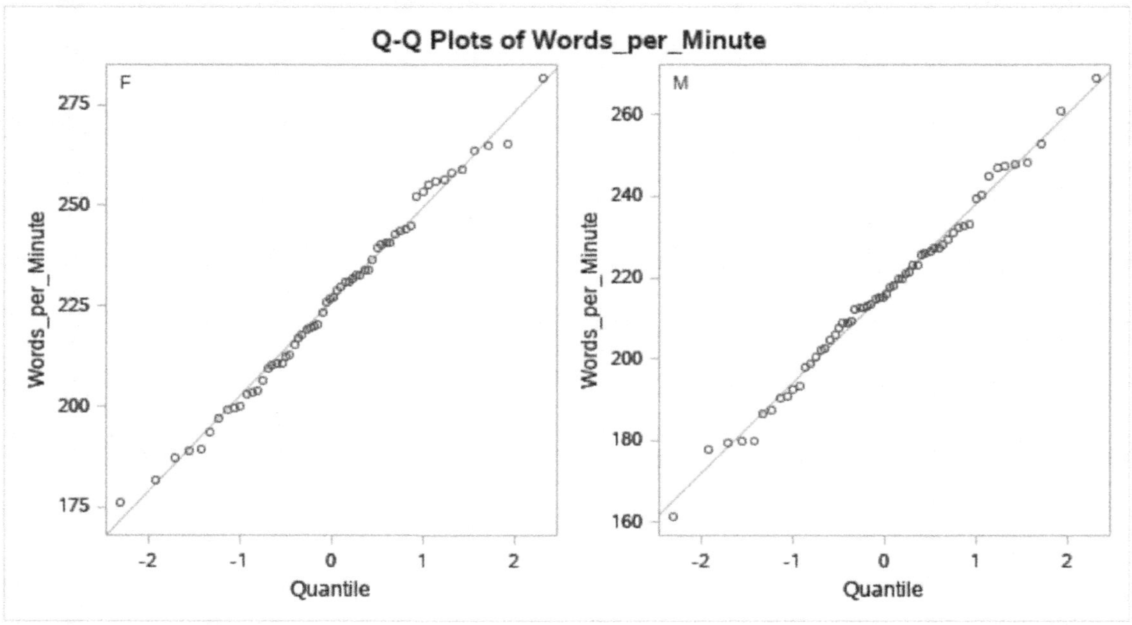

Figure 8.12: Q-Q Plots for Females and Males

Nonparametric Two-Sample Tests

If you want to compare two groups and you feel that the assumptions for a parametric test are not satisfied, the *t* Tests statistics task can also perform a **Wilcoxon Rank Sum test**, the nonparametric counterpart to the Student's *t* test.

To demonstrate how this works, let's use the SASHELP data set called Fish to demonstrate the Wilcoxon test. The Fish data set contains weights and several other variables for several different species of fish. You decide to compare weights of two species: Roach and Pike. (Note: This author has never heard of a Roach fish.)

This example has two purposes: To demonstrate how to run and interpret a Wilcoxon Rank Sum test and to review how to filter rows in a table.

You start by double-clicking the *t* Tests task as before. On the DATA tab, click Filter and write an expression that selects only Roach and Pike. On this same tab, select a two-sample test, select Weight as your analysis variable and Species as your Groups variable. (See Figure 8.13.)

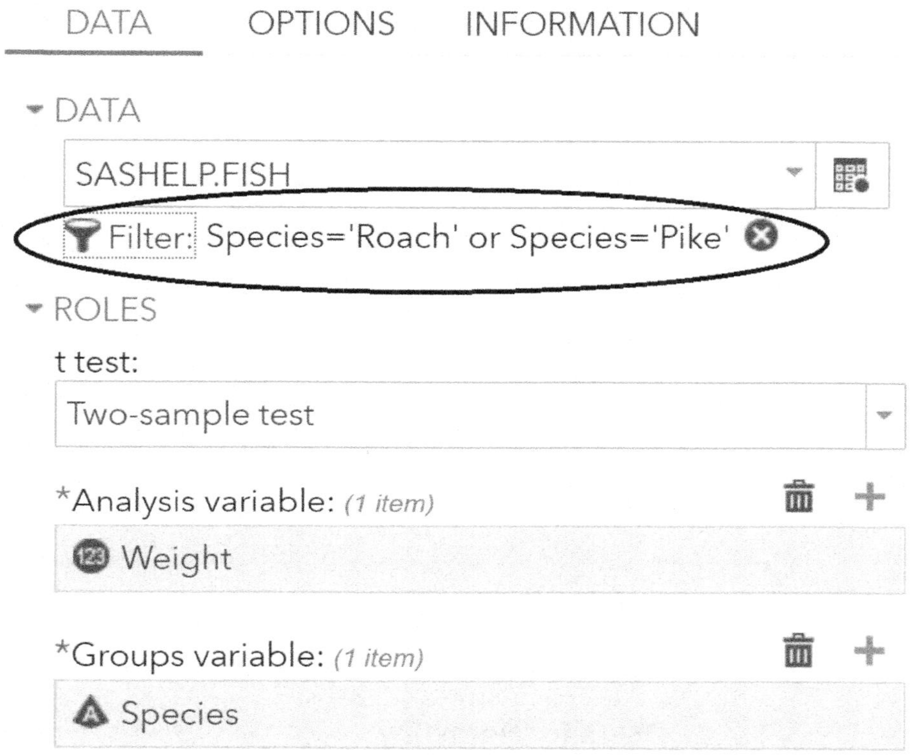

Figure 8.13: Requesting a Filter and a Two-Sample Test

> Because Species is a character variable, the values "Pike" and "Roach" must be in single or double quotation marks in the filter expression. For numeric variables, do not place quotation marks around the numeric values.

You can undo a filter by clicking the small 'x' next to the filter text in the DATA window.

You can now click the Run icon to display the results for the test of normality and see histograms for weights of the two selected fish species. The normality test results are shown in Figure 8.14, and the histograms are displayed in Figure 8.15.

Variable: Weight
Species = Pike

Tests for Normality				
Test	**Statistic**		**p Value**	
Shapiro-Wilk	**W**	0.821287	**Pr < W**	0.0040
Kolmogorov-Smirnov	**D**	0.267641	**Pr > D**	<0.0100
Cramer-von Mises	**W-Sq**	0.218639	**Pr > W-Sq**	<0.0050
Anderson-Darling	**A-Sq**	1.255384	**Pr > A-Sq**	<0.0050

Variable: Weight
Species = Roach

Tests for Normality				
Test	**Statistic**		**p Value**	
Shapiro-Wilk	**W**	0.932277	**Pr < W**	0.1708
Kolmogorov-Smirnov	**D**	0.176514	**Pr > D**	0.0986
Cramer-von Mises	**W-Sq**	0.102833	**Pr > W-Sq**	0.0968
Anderson-Darling	**A-Sq**	0.572359	**Pr > A-Sq**	0.1243

Figure 8.14: Test of Normality for Two Species

All of the tests for normality for Pike reject the null hypothesis that the weights are normally distributed. The histograms and box plots shown in Figure 8.15 make it clear that the distribution for Pike deviates considerably from a normal distribution. The plots are shown below:

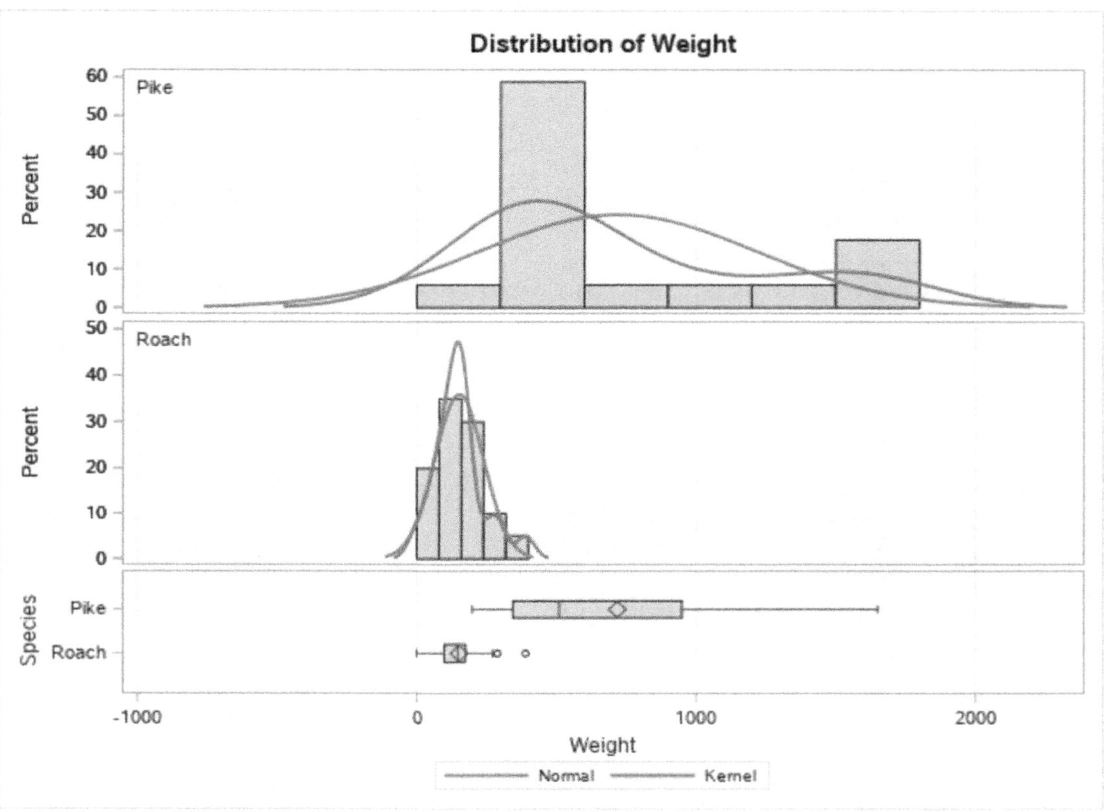

Figure 8.15: Distributions for Fish Weights

You need to pay closer attention to the distributions of weights because of the much smaller sample sizes for these two fish species (n=20 for Roach and n=17 for Pike).

Because of these results, you decide to run a Wilcoxon Rank Sum test. To do this, simply check the box labeled Wilcoxon Rank Sum Test on the OPTIONS tab, as shown in Figure 8.16.

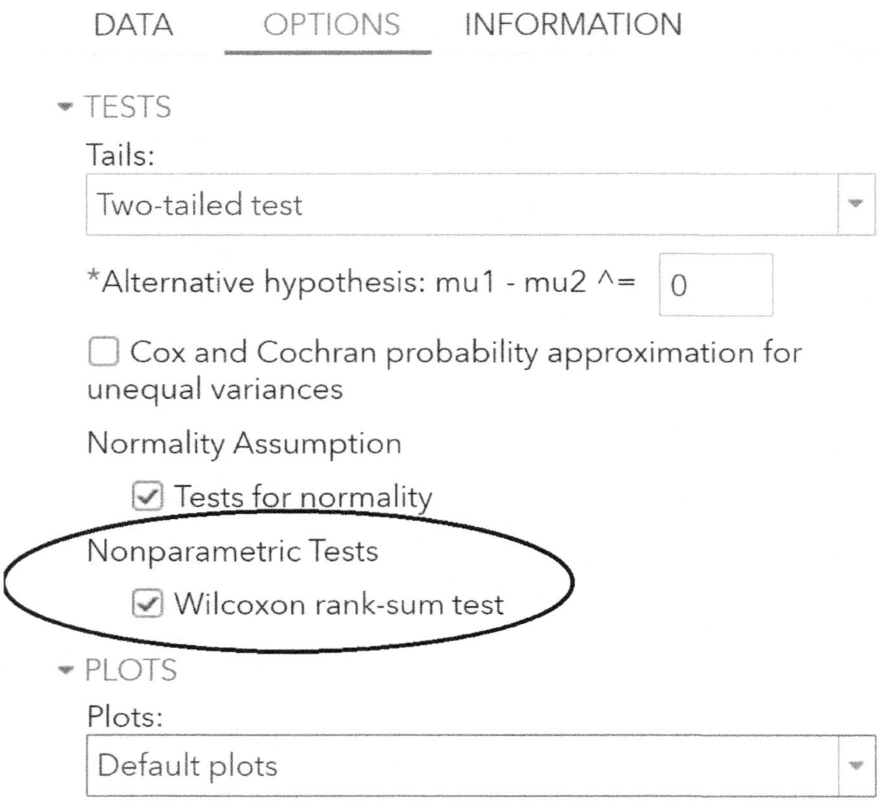

DATA OPTIONS INFORMATION

TESTS

Tails:

Two-tailed test

*Alternative hypothesis: mu1 - mu2 ^= 0

☐ Cox and Cochran probability approximation for unequal variances

Normality Assumption

☑ Tests for normality

Nonparametric Tests

☑ Wilcoxon rank-sum test

PLOTS

Plots:

Default plots

Figure 8.16: Requesting the Wilcoxon Rank Sum Test

Because you have already seen the histograms, you can use the menu under Plots to select a Wilcoxon box plot and deselect the other plots. (See Figure 8.17.)

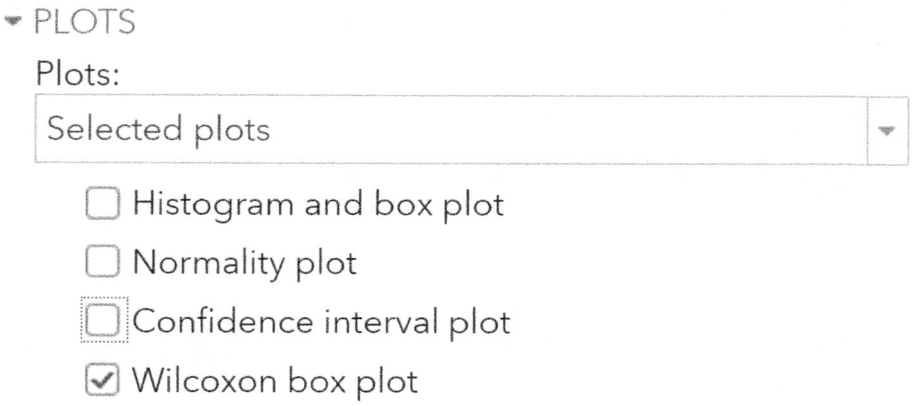

▾ PLOTS

Plots:

Selected plots ▾

☐ Histogram and box plot

☐ Normality plot

☐ Confidence interval plot

☑ Wilcoxon box plot

Figure 8.17: Requesting a Wilcoxon Box Plot

Click the Run icon to obtain the following:

Wilcoxon Scores (Rank Sums) for Variable Weight Classified by Variable Species					
Species	**N**	**Sum of Scores**	**Expected Under H0**	**Std Dev Under H0**	**Mean Score**
Roach	20	217.50	380.0	32.798983	10.875000
Pike	17	485.50	323.0	32.798983	28.558824
Average scores were used for ties.					

Figure 8.18: Rank Sums for the Wilcoxon Rank Sum Test

Figure 8.18 shows the sum of ranks for the two fish species. Here's how it works: You order all the fish weights (both species combined), from lowest to highest, giving them ranks (the smallest is rank 1, the next smallest is rank 2, and so on). If there are two or more weights that are equal, you give them all an average rank. Finally, you add up the ranks for each species. In the figure above, you see this sum labeled Sum of Scores. If the null hypothesis is true, you would expect the sum of ranks to be about the same for both groups (depending on the sample sizes). Looking at the two histograms of weights, you see that, in general, the weights of Roach fish are lighter than the weights of Pike. Therefore, you expect the sum of ranks for Roach to be lower than the sum of ranks for Pike. If the null hypothesis is true, large differences between the sum of ranks in the two samples are unlikely. That is the basis for computing a *p*-value for this test.

The next portion of the output shows you two ways to compute a *p*-value. One, useful for larger samples, is a z-test (with a correction for continuity); the other test, sometimes used for smaller samples, is a *t* approximation. In this table, both *p*-values are very small, and you can claim that Pike are heavier than Roach based on this test.

<table>
<tr><th colspan="6">Wilcoxon Two-Sample Test</th></tr>
<tr><th rowspan="2">Statistic</th><th rowspan="2">Z</th><th rowspan="2">Pr > Z</th><th rowspan="2">Pr > |Z|</th><th colspan="2">t Approximation</th></tr>
<tr><th>Pr > Z</th><th>Pr > |Z|</th></tr>
<tr><td>485.5000</td><td>4.9392</td><td><.0001</td><td><.0001</td><td><.0001</td><td><.0001</td></tr>
<tr><td colspan="6">Z includes a continuity correction of 0.5.</td></tr>
</table>

Figure 8.19: p-Values for the Wilcoxon Test

Because you checked the box for a Wilcoxon box plot, you are presented with Figure 8.20.

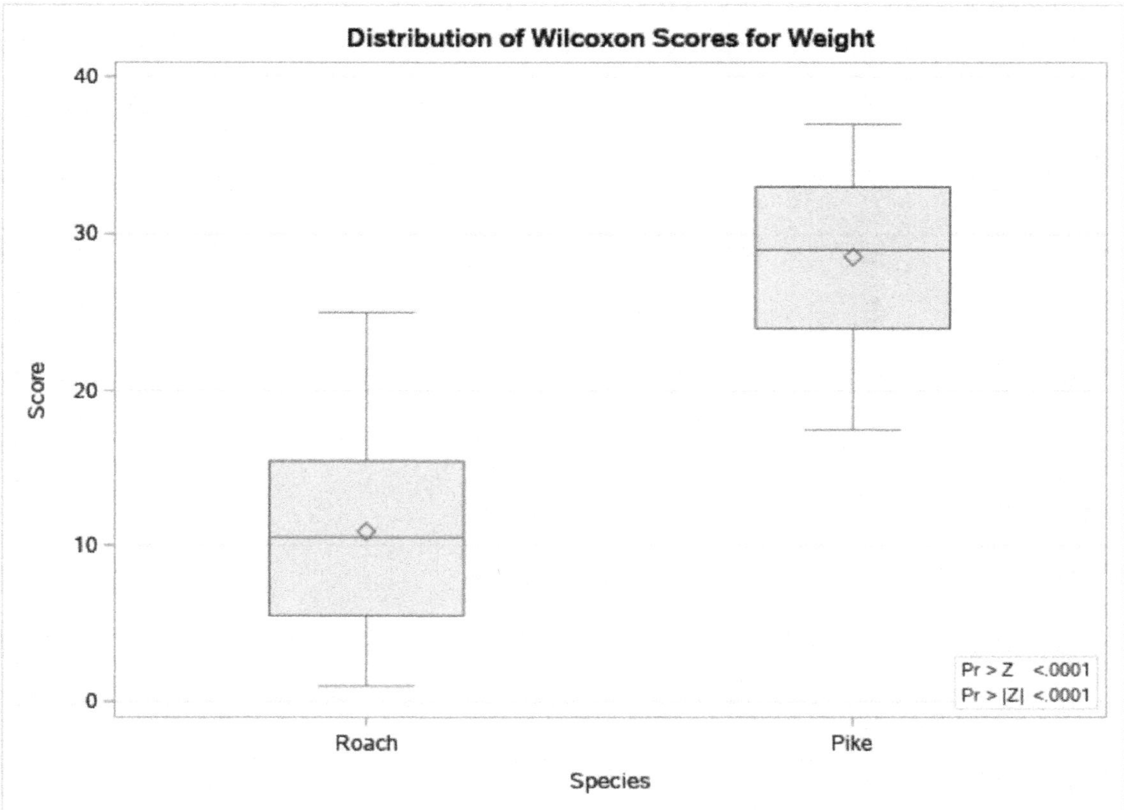

Figure 8.20: Distribution of Wilcoxon Scores

This plot shows the distribution of ranks for the two groups of fish. You can see that the Pike weights are given most of the larger rank values.

Paired t Test

The final section of this chapter describes a **paired *t* test**. This test is used when the situation either involves two values taken from each subject (such as a score taken before and after a subject is treated) or two subjects who are paired on one or more characteristics. Then one is randomly assigned to one treatment and the second to the other treatment

To demonstrate a paired *t* test, you can use data from a small study designed to show if a half hour of yoga can lower a subject's heart rate. Ten subjects had their heart rate measured before and after a yoga session, and the results were entered into an Excel workbook called Yoga.xlsx. This workbook was saved in the folder Statistics Data. Figure 8.21 shows the original spreadsheet.

	A	B	C
1	Subj	Before	After
2	1	78	74
3	2	68	68
4	3	76	70
5	4	58	57
6	5	83	73
7	6	80	77
8	7	69	61
9	8	77	76
10	9	77	72
11			

Figure 8.21: Spreadsheet Containing Before and After Heart Rates

The Import Utility under the Tasks and Utilities tab was used to convert the workbook into a SAS data set called Yoga that was placed in the WORK library. The next step is to double-click the *t* Tests tab and enter the appropriate information on the DATA tab. The data set name is WORK.Yoga. A paired *t* test is selected from the menu, and the two variables, Before and After, are entered as the Group 1 and Group 2 variables. (See Figure 8.22 below).

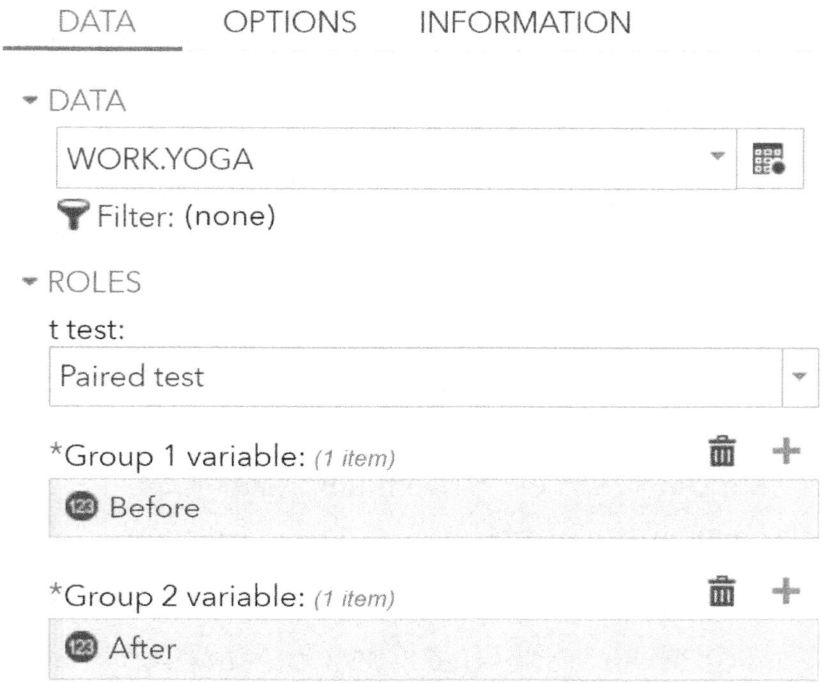

DATA OPTIONS INFORMATION

DATA

WORK.YOGA

Filter: (none)

ROLES

t test:

Paired test

*Group 1 variable: *(1 item)*

Before

*Group 2 variable: *(1 item)*

After

Figure 8.22: Data Tab for Paired t Test

It's time to run the procedure (you are leaving all the defaults on the OPTIONS tab). Here are the results:

Variable: _Difference_ (Difference: Before - After)

Tests for Normality				
Test	**Statistic**		**p Value**	
Shapiro-Wilk	**W**	0.952338	**Pr < W**	0.7157
Kolmogorov-Smirnov	**D**	0.162908	**Pr > D**	>0.1500
Cramer-von Mises	**W-Sq**	0.027145	**Pr > W-Sq**	>0.2500
Anderson-Darling	**A-Sq**	0.209771	**Pr > A-Sq**	>0.2500

Figure 8.23: Test for Normality for Difference Scores

All of the test for normality are not significant. You should not interpret this to mean that the difference scores are normally distributed. With such a small sample (10 subjects), it would take large deviations from a normal distribution to reject the null hypothesis at the .05 level. You will need to look at the histogram (or a Q-Q plot) to help decide if a parametric test is reasonable.

The next section of the output shows the *t* table. You see that the mean difference is 4.222, the *t* value is 3.74, and the *p*-value is .0057. Because the mean difference is positive and the difference score was computed as the before value minus the after value, you can conclude that the yoga session helped reduce heart rate (at the .05 level).

Difference: Before - After

N	Mean	Std Dev	Std Err	Minimum	Maximum
9	4.2222	3.3830	1.1277	0	10.0000

Mean	95% CL Mean		Std Dev	95% CL Std Dev	
4.2222	1.6218	6.8226	3.3830	2.2850	6.4810

DF	t Value	Pr > |t|
8	3.74	0.0057

Figure 8.24: Statistics, t- and p-Values

The next figure shows a histogram for the difference scores. Although it doesn't look too much like a normal distribution, it is fairly symmetrical. With a sample size of 10, you decide that a *t* test is appropriate.

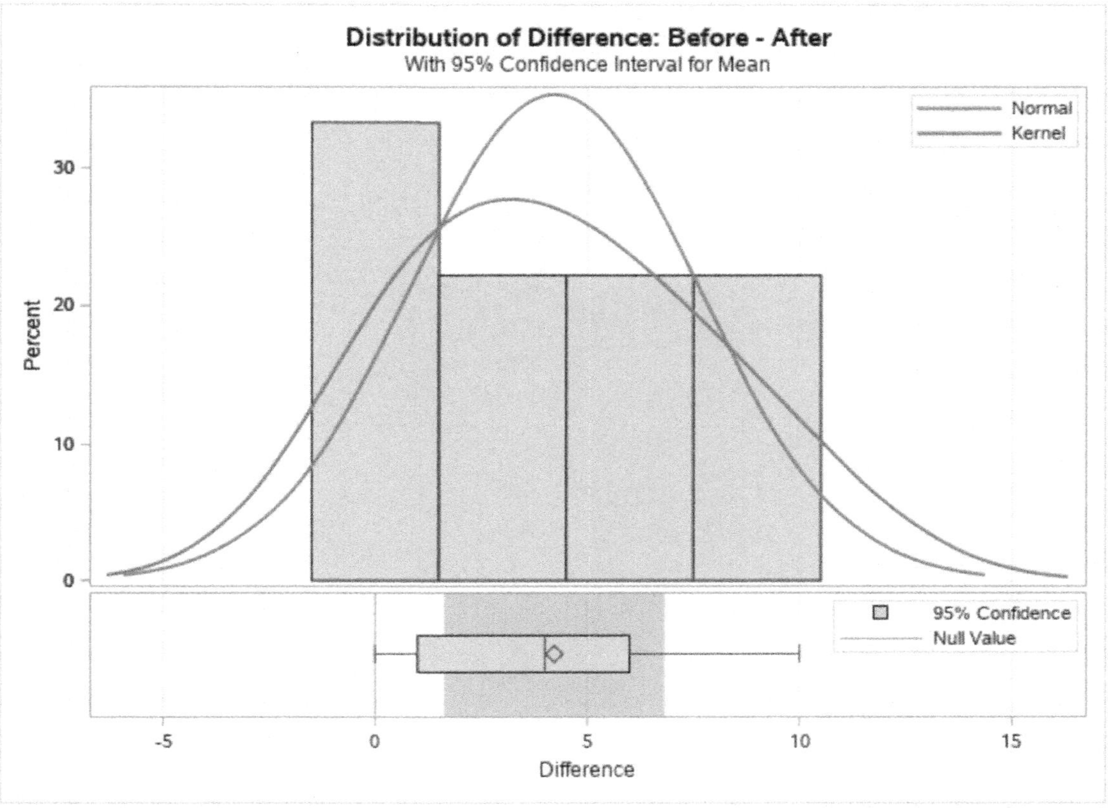

Figure 8.25: Histogram of Difference Scores

If you have any doubts, you should rerun the analysis using a nonparametric test such as the Wilcoxon Signed Rank test. This is accomplished by checking the box on the OPTIONS tab that requests this test. (Actually, the option runs both a Sign test and a Wilcoxon Signed Rank test.) Although not shown here, these two nonparametric tests both show a significant difference at the .05 level.

There are several more plots in the output, and they will not all be shown. However, one of the plots, Paired Profiles, shows the Before and After score for each person in the study as well as the mean scores for before and after. It is shown below.

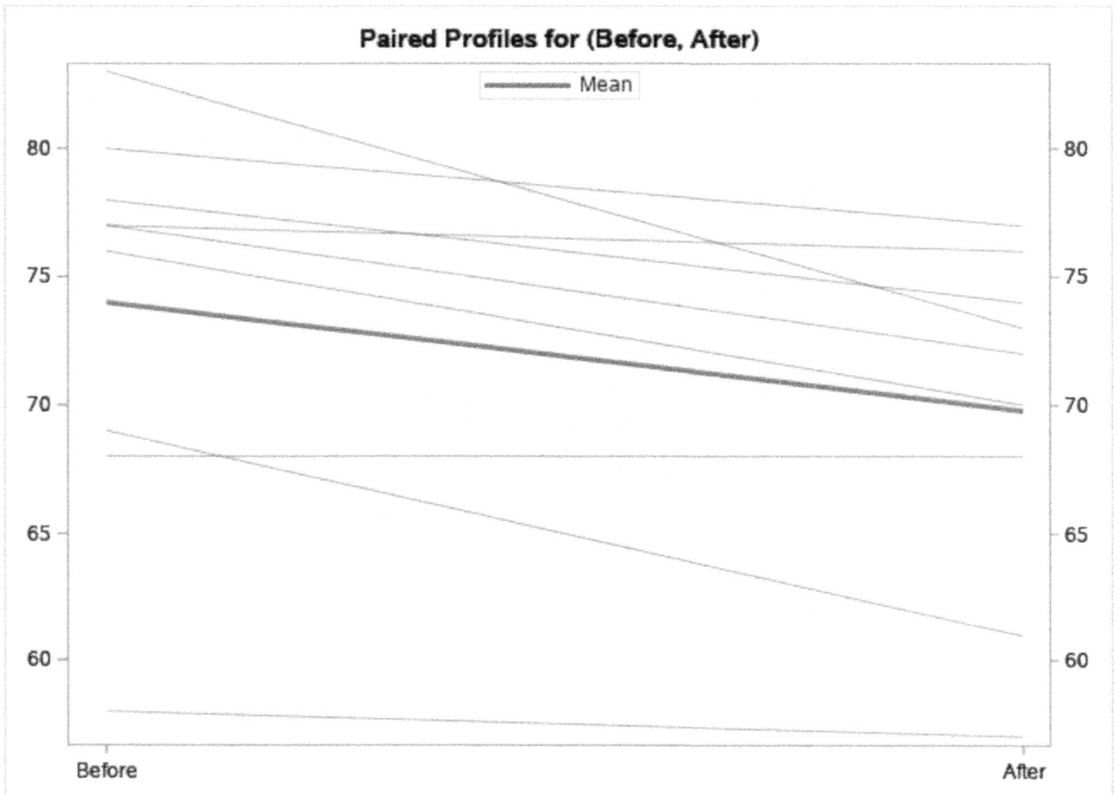

Figure 8.26: Paired Profiles Plot

You can see that 9 out of 10 subjects had a lower heart rate after the yoga exercise.

Conclusion

You have seen that the *t* Tests statistics task can perform one-sample *t* tests as well as two-sample paired and unpaired tests. For each of these tests, one or more nonparametric alternatives are provided. Because one of the assumptions for all of the *t* tests is that the data values are normally distributed (or close to it, depending on the sample size), the task provides you with several tests of normality as well as histograms and Q-Q plots.

Chapter 8 Exercises

> Several of the exercises in this chapter and several in subsequent chapters use permanent SAS data sets created when you run the program Create_Datasets.sas (in the download package). To save you the trouble of rerunning this program each time you start a new session, add the line: libname Stats '~/Statistics Data'; (that is the folder name I use in the book, change it if you want) to a file called Autoexec.sas. You can edit this file by clicking on the icon shown below this box. Choose Edit Autoexec File. This will set up the Stats library every time you start a SAS Studio session.

1. Using the Heart data set in the SASHELP library, compare Systolic (systolic blood pressure) for males and females (variable Sex with values of F and M). Conduct the tests for normality. Do you reject the null hypothesis that the data values are normally distributed? Do you think that is important?
2. Using the data set High_School in the STATS library, conduct a two-sample *t* test to determine whether the two variables English_Grade and Vocab_Score are significantly different based on Gender.
3. (Advanced) Enter data into a spreadsheet called Ttest_Data.xlsx (or xls) so that it looks like the image below:

	A	B	C
1	Method	Score1	Score2
2	A	96	96
3	A	94	97
4	A	98	93
5	A	93	94
6	B	89	91
7	B	88	88
8	B	85	96
9	B	84	95

If you are using SAS Studio with ODA, place this file in the Statistics Data folder (or whatever name you choose). Next, use the import facility to import this file into a temporary (WORK) data set called Ttest_Data. Run a *t* test on Score1 and Score2, using the variable Method as the grouping variable. For Score2, also run a Wilcoxon Rank Sum Test.

4. Run a two-sample *t* test comparing the weight of Roach and Parkki in the SASHELP.FISH data set. Do this using both parametric and nonparametric methods. Be sure to generate a histogram and box plot to investigate the distribution of Weight for the two species. You will need to create a filter (Species='Roach' or Species='Parkki').

5. Using the data set Cars in the SASHELP library, compare the invoice amount (variable Invoice) for four- versus six-cylinder cars. You will need to create a filter to restrict the variable Cylinders for values of 4 or 6. (Hint: In the filter box, enter: Cylinders=4 or Cylinders=6.) Notice that the tests for normality are all highly significant. However, there are 136 four-cylinder cars and 190 six-cylinder cars. Just to be sure, run a Wilcoxon Rank Sum test to see if you get the same result.

6. Run the short program below and conduct a two-way test with X as the Analysis variable and Group as the Groups variable. If you inspect the program, you will get an idea of how simulated data sets were created for this book. For the curious reader, the RAND function in this program is generating normally distributed random numbers with a mean of 100 and a standard deviation of 15. If the subject is in Group B, the program adds 10 to the value of X.

```
data TTest;
   call streaminit(13579);
   do Subj = 1 to 10;
      Do Group = 'A','B';
         X = round(rand('normal',100,15) + 10*(Group = 'B'));
         output;
      end;
   end;
run;
```

7. Repeat exercise 6, except change the 10 in the DO loop to 50. Rerun the *t* test and compare the *p*-value to what you obtained in exercise 6.

9.

Comparing More Than Two Means (ANOVA)

Introduction

When you want to compare means in a study where there are three or more groups, you cannot use multiple *t* tests. In the old days (even before my time!), if you had three groups (let's call them A, B, and C), you might perform *t* tests between each pair of means (A versus B, A versus C, and B versus C). With four groups, the situation gets more complicated; you would need six *t* tests (A versus B, A versus C, A versus D, B versus C, B versus D, and C versus D). Even though no one does multiple *t* tests anymore, it is important to understand the underlying reason why this is not statistically sound.

Suppose you are comparing four groups and performing six *t* tests. Also, suppose that the null hypothesis is true, and all the means come from populations with equal means. If you perform each *t* test with α set at .05, there is a probability of .95 that you will make the correct decision—that is, to fail to reject the null hypothesis in each of the six tests. However, what is the probability that you will reject at least one of the six null hypotheses? To spare you the math, the answer is about .26 (or 26% if that is easier to think about). This is called an "experimentwise" type I error. Remember, a type I error is when you reject the null hypothesis (claim the samples come from populations with different means—"the drug works") when you shouldn't. So, instead of your chance of reporting a false positive result being .05, it is really .26.

To prevent this problem, statisticians came up with a single test, called **analysis of variance** (abbreviated ANOVA). The null hypothesis is that all the means come from populations with equal means; the alternative hypothesis is that there is at least one mean that is different from the others. You either reject or fail to reject the null hypothesis, and there is one *p*-value associated with the test. If you reject the null hypothesis, you can then investigate pairwise differences using methods that control the experimentwise type I error.

Getting an Intuitive Feel for a One-Way ANOVA

Before we get into the details of running and interpreting ANOVA tables, let's get an intuitive feel for how this analysis works. Suppose you have three groups of subjects (A, B, and C) and you collected the following data:

Group	A	B	C
	50	78	20
	45	80	15
	55	82	26
Means	**50**	**80**	**20**

You see the means in groups A, B, and C are 50, 80, and 20 respectively. They seem pretty far apart. But, what does "far apart" mean? In this case, they are far apart compared to the scores within each group (which seem very close to the group mean). This might lead you to think that there is a significant difference between the groups.

> In English, when there are more than two groups, proper grammar is to say "among", not "between". However, the terms "within" and "between" have been used to describe variances in ANOVA designs since they were first developed, and many textbooks have kept with these terms.

You can skip this next paragraph if you want—it describes how ANOVA works in more detail.

You can estimate the population variance by looking at the scores within a group or by using the group means to estimate the variance. If the null hypothesis were true (all the sample means come from populations with equal means), these two estimates of variance would be about the same and the ratio of the between-group variance to the within-group variance (called an **F value**) would be close to 1. If there were significant differences between the groups, the variance estimate computed by using the group means would be larger than the variance estimate computed by looking at the scores within a group. In this case, the F ratio would be greater than 1.

Performing a One-Way Analysis of Variance

We can use the data set called Reading (in the STATS library), containing data on reading speeds of males and females, as well as three different methods that might improve reading speeds of the test subjects, to demonstrate a one-way ANOVA.

Start by choosing the task One-Way ANOVA from the Statistics tasks Linear Models task list. This brings up the following screen:

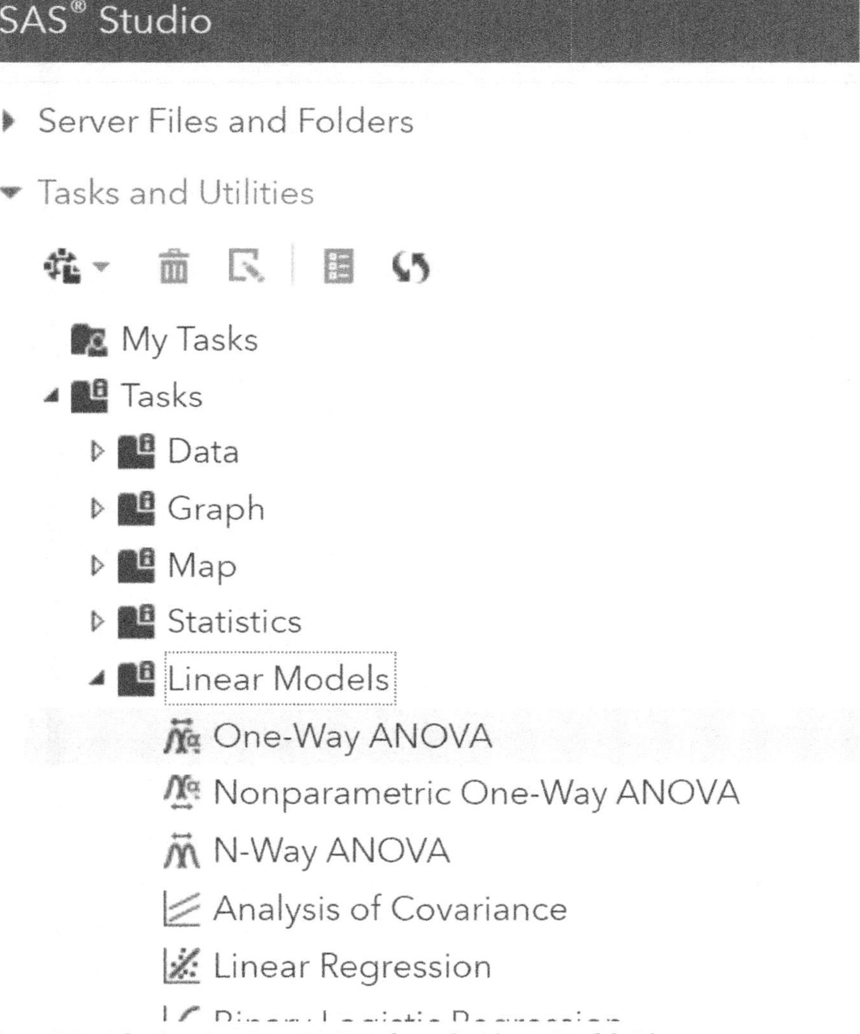

Figure 9.1: Selecting One-Way ANOVA from the Linear Models List

To begin, double-click the One-Way ANOVA task tab.

Choose the data set Reading in the STATS library. This library was created when you ran the program Create_Datasets.sas. Choose Words_per_Minute as the Dependent variable and Method as the Categorical (independent) variable, as shown in Figure 9.2.

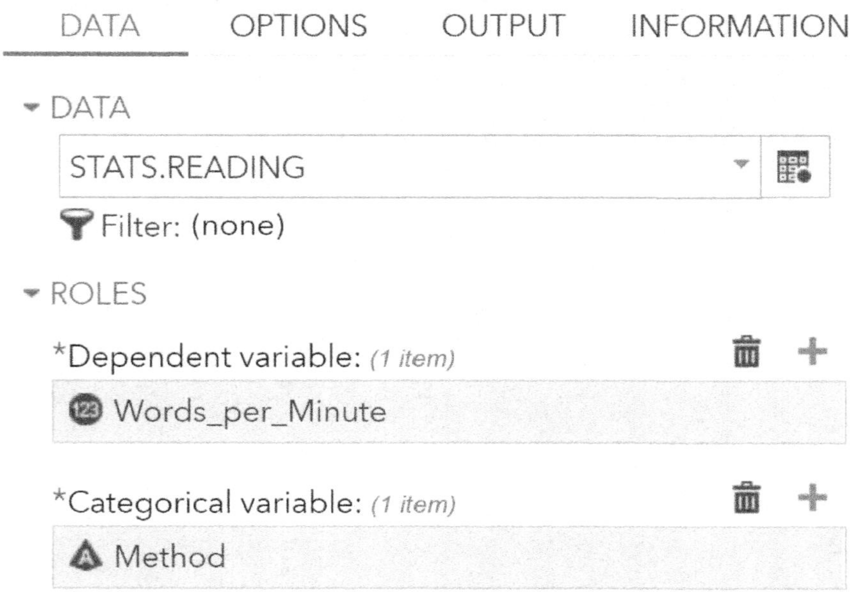

DATA OPTIONS OUTPUT INFORMATION

DATA

STATS.READING

Filter: (none)

ROLES

*Dependent variable: (1 item)

Words_per_Minute

*Categorical variable: (1 item)

Method

Figure 9.2: DATA Tab for One-Way ANOVA

Once you have completed the DATA screen, click the OPTIONS tab to see the following.

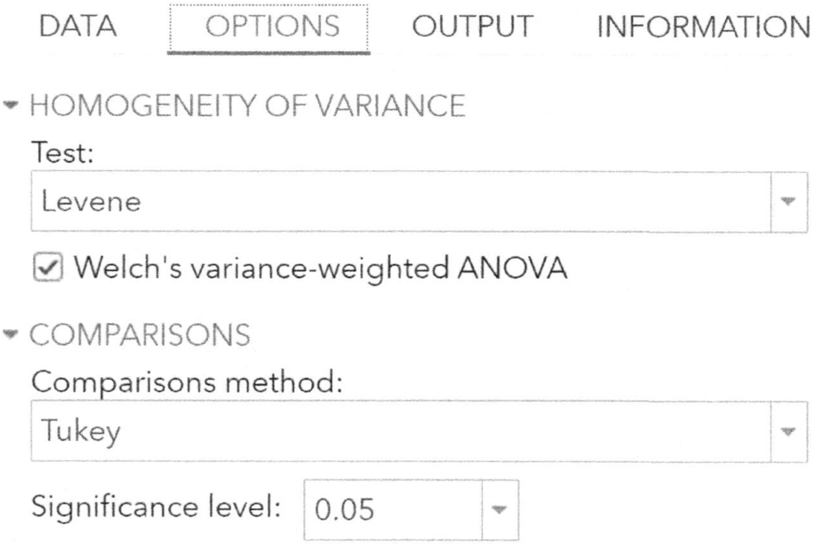

DATA OPTIONS OUTPUT INFORMATION

HOMOGENEITY OF VARIANCE

Test:

Levene

Welch's variance-weighted ANOVA

COMPARISONS

Comparisons method:

Tukey

Significance level: 0.05

Figure 9.3: OPTIONS for One-Way ANOVA (Top Portion)

One of the assumptions for performing an analysis of variance is that the variances in each of the groups are equal. The **Levene test** is used to determine if this assumption is reasonable. If this test is significant (meaning that the variances are not equal), you might choose to ignore it if the differences are not too large. (ANOVA is said to be robust to the assumption of equal variance, especially if the sample sizes are similar). If you want to account for unequal variances, click the box for Welch's variance-weighted ANOVA.

Multiple comparisons are methods that you use in order to determine which pairs of means are significantly different. There are several choices for these tests. The default is Tukey, a popular choice. Later in this chapter, you will see another multiple comparison test called SNK (Student-Newman-Keuls). You probably want to leave the significance level at .05.

It's time to run the procedure. Click the Run icon to produce the tables and graphs.

The first section of output displays class-level information. Don't ignore this! Make sure that the number of levels is what you expected (data errors can cause the program to believe there are more levels than there are). Also, pay attention to the number of observations read and used. This is important because any missing values on either the dependent variable (Words_per_Minute) or categorical variable (Method) will result in that observation being omitted from the analysis. A large proportion of missing values in the analysis can lead to bias—subjects with missing values might be different in some way from subjects without missing values (that is, missing values might not be random).

Class Level Information		
Class	Levels	Values
Method	3	A B C

Number of Observations Read	120
Number of Observations Used	120

Figure 9.4: Class-Level Information

There are three levels for Method (A, B, and C) and there are no missing values (because the number of observations read is the same as the number of observations used). It's time to look at your ANOVA table (Figure 9.5 below).

Dependent Variable: Words_per_Minute

Source	DF	Sum of Squares	Mean Square	F Value	Pr > F
Model	2	16856.53834	8428.26917	20.84	<.0001
Error	117	47324.92840	404.48657		
Corrected Total	119	64181.46674			

Figure 9.5: ANOVA Table

You can look at the F test and *p*-values in the ANOVA table, but you must remember that you also need to look at several other parts of the output to determine if the assumptions for the test are satisfied. You will see in the diagnostic tests that follow that the ANOVA assumptions were satisfied, so let's go ahead and see what conclusions you can draw from the ANOVA table and the tables that follow.

Notice that the model has 2 **degrees of freedom** (because there were 3 levels of the independent variable, and the degrees of freedom is the number of groups minus 1). The mean squares for the model and error terms tell you the between-group variance and the within-group variance. The ratio of these two variances, the F value, is 20.84 with a corresponding *p*-value of less than .0001. A result such as this is often referred to as "highly significant." Remember, the term "significant" means that the probability of falsely rejecting the null hypothesis is smaller than a pre-determined value. It doesn't necessarily mean that the differences are significant in the common usage of the word, that is, important.

To graphically display the distribution of reading speed (Words_per_Minute) in the three groups, the one-way ANOVA task produces a box plot (Figure 9.6). The line in the center of the box represents the median, and the small diamond represents the mean. Notice that the means, as well as the medians, of the three groups are not very different. Why then were the results so highly significant? The reason is the large sample size (120). Large sample sizes give you high power to see even small differences.

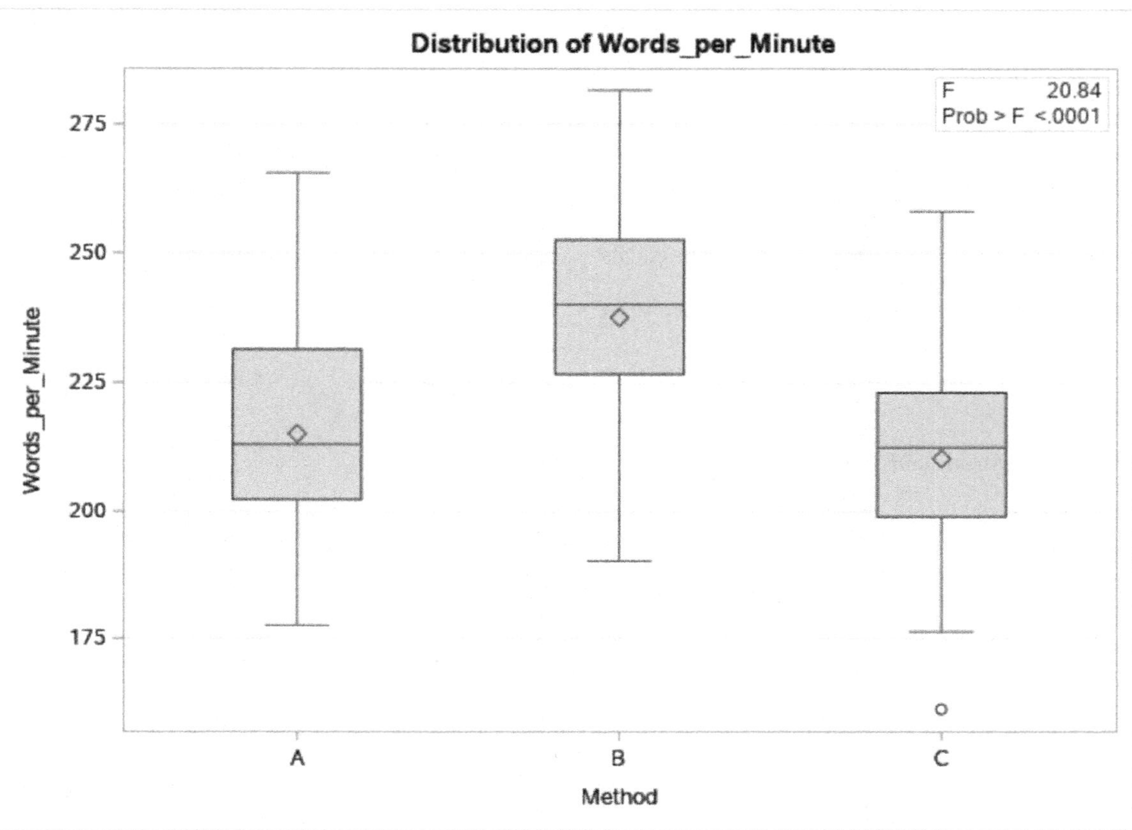

Figure 9.6: Box Plot for Words_per_Minute by Method

Figure 9.7 shows the results for Levin's test of homogeneity of variance. Here, the null hypothesis is that the variances are equal. Because the *p*-value is .9425, you do not reject the null hypothesis of equal variance.

Levene's Test for Homogeneity of Words_per_Minute Variance ANOVA of Squared Deviations from Group Means					
Source	**DF**	**Sum of Squares**	**Mean Square**	**F Value**	**Pr > F**
Method	2	36892.2	18446.1	0.06	0.9425
Error	117	36445775	311502		

Figure 9.7: Levene's Test for Homogeneity of Variance

Figure 9.8 show the means and standard deviations for the three groups.

Level of Method	N	Words_per_Minute	
		Mean	Std Dev
A	40	215.161350	20.7279745
B	40	237.627783	19.7155850
C	40	210.470850	19.8772856

Figure 9.8: Means and Standard Deviations for the Three Groups

Because this is a one-way model, the least square means shown in Figure 9.9 are equal to the means computed by adding up all the values within a group and dividing by the number of subjects in that group. In unbalanced models with more than one factor, this might not be the case.

Below the table showing the three means, you see *p*-values for all of the pairwise differences. Each of the three reading methods in the top table in the figure has what is labeled as the LSMEAN Number. In the table of *p*-values, the LSMEAN number is used to identify the groups. The intersection of any two groups displays the *p*-value for the difference. For example, group 1 (Method A) and group 2 (Method B) show a *p*-value of less than .0001. The *p*-value for the difference of Method A (1) and Method C (3) is .5514 (not significant).

Least Squares Means
Adjustment for Multiple Comparisons: Tukey

Method	Words_per_Minute LSMEAN	LSMEAN Number
A	215.161350	1
B	237.627783	2
C	210.470850	3

Least Squares Means for effect Method
Pr > |t| for H0: LSMean(i)=LSMean(j)

Dependent Variable: Words_per_Minute

i/j	1	2	3
1		<.0001	0.5514
2	<.0001		<.0001
3	0.5514	<.0001	

Figure 9.9: Least Square Means

Figure 9.10 shows a very clever way to display pairwise differences. At the intersection of any two groups, you see a diagonal line representing a 95% confidence interval for the difference between the two group means. If the interval crosses the main diagonal line (that represents no difference), the two group means are not significantly different at the .05 level. To make this clearer, significant differences are seen in the two top diagonal lines representing C versus B and A versus B (they don't cross the dotted line) and the diagonal line at the bottom left of the diagram representing C versus A, indicates a non-significant difference. By the way, the name **diffogram** is used to describe this method of displaying pairwise differences.

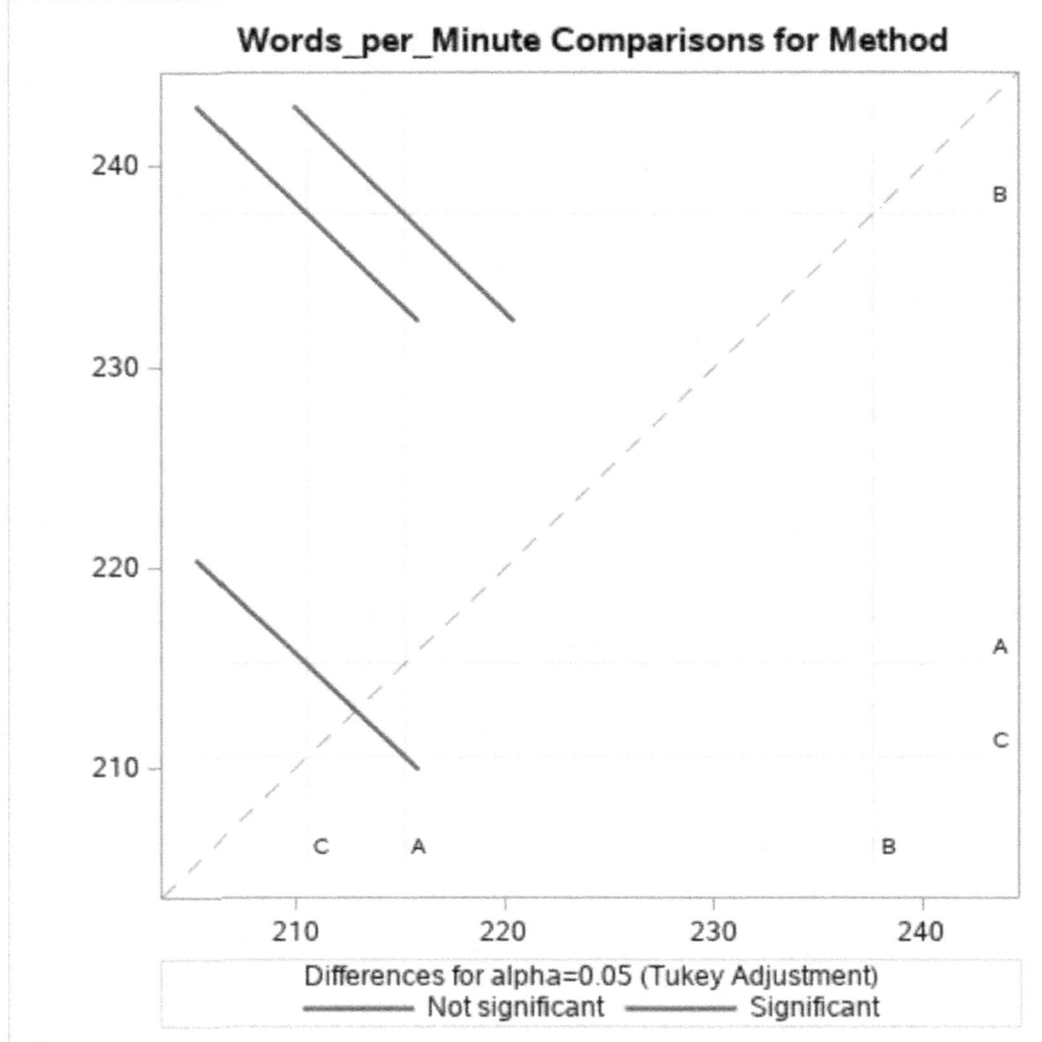

Figure 9.10: Pairwise Comparison of Means

All of the previous figures were generated by the choices that you made in the DATA and OPTIONS tabs. There is an alternative method of determining pairwise differences called the **Student-Newman-Keuls (SNK) test** (also referred to in some texts as just Newman-Keuls). The SNK test is similar to the Tukey test in that it shows group means and which pairs of means are different at the .05 level. The Tukey test has the advantage of computing *p*-values for each pair of means as well as a confidence interval for the differences. The SNK test can do neither of these two things but has a slightly higher power to detect differences.

To request the SNK multiple comparison test, select Student-Newman-Keuls from the list under the Comparisons tab as shown in Figure 9.11.

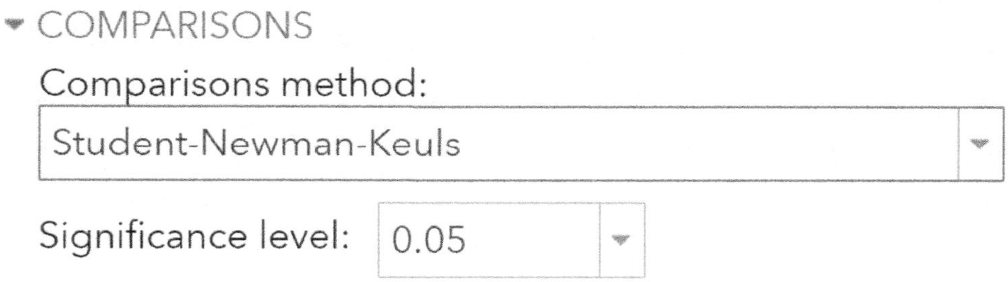

▾ COMPARISONS

Comparisons method:

Student-Newman-Keuls

Significance level: 0.05

Figure 9.11: Requesting an SNK Multiple Comparison Test

The SNK display (Figure 9.12) shows the three means in order from highest to lowest. Notice the bars on the right side of the output. Any two means that share the same bar are not significantly different at the .05 level. You can see here that the mean reading speed for group B is the highest, and it is significantly different from the mean of group A and from group C. Because groups A and C share a single bar, these two means (215.16 and 210.47) are not significantly different from each other.

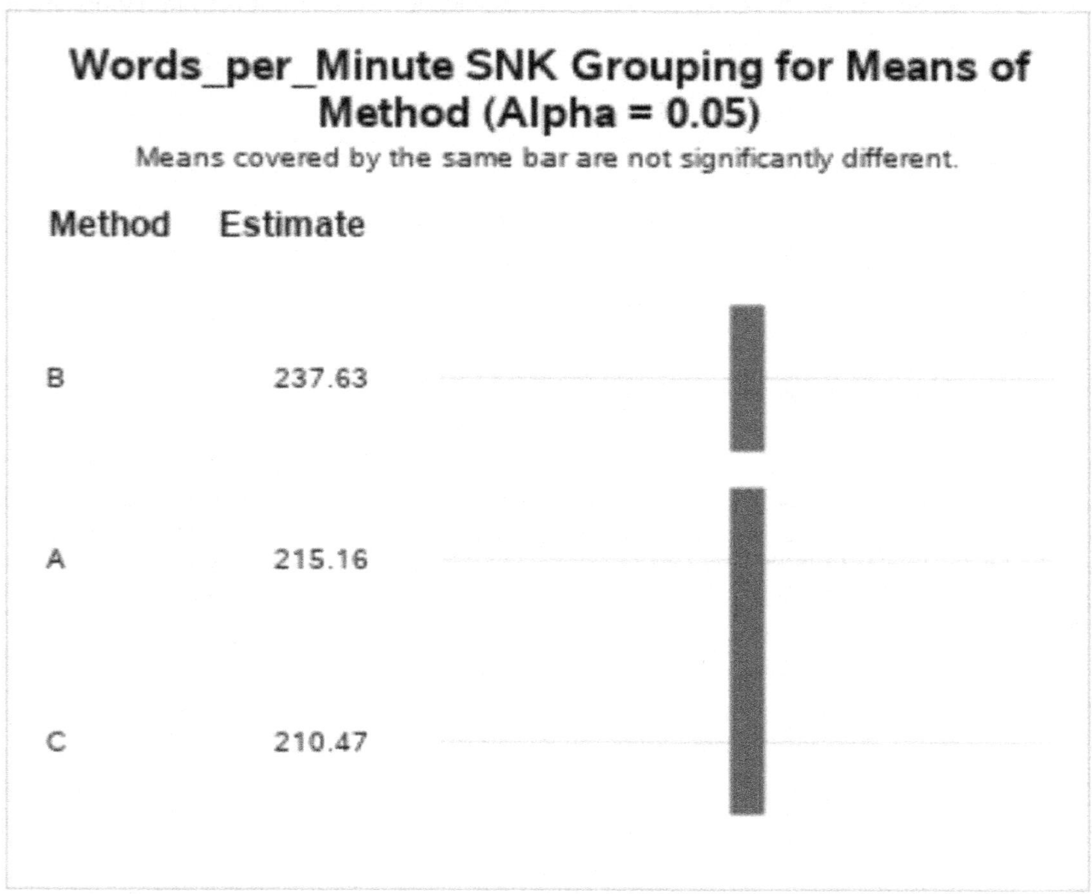

Figure 9.12: Student-Newman-Keuls Pairwise Comparisons

Performing More Diagnostic Plots

Before we leave this section, let's look at a few diagnostic plots that you can choose on the PLOTS menu. In the pull-down list below Diagnostic Plots, you can select either Panel of Plots or Individual plots. The Panel option shows all the plots on a single page—the Individual option shows each diagnostic plot on a separate page. In this example, you decided to see individual plots.

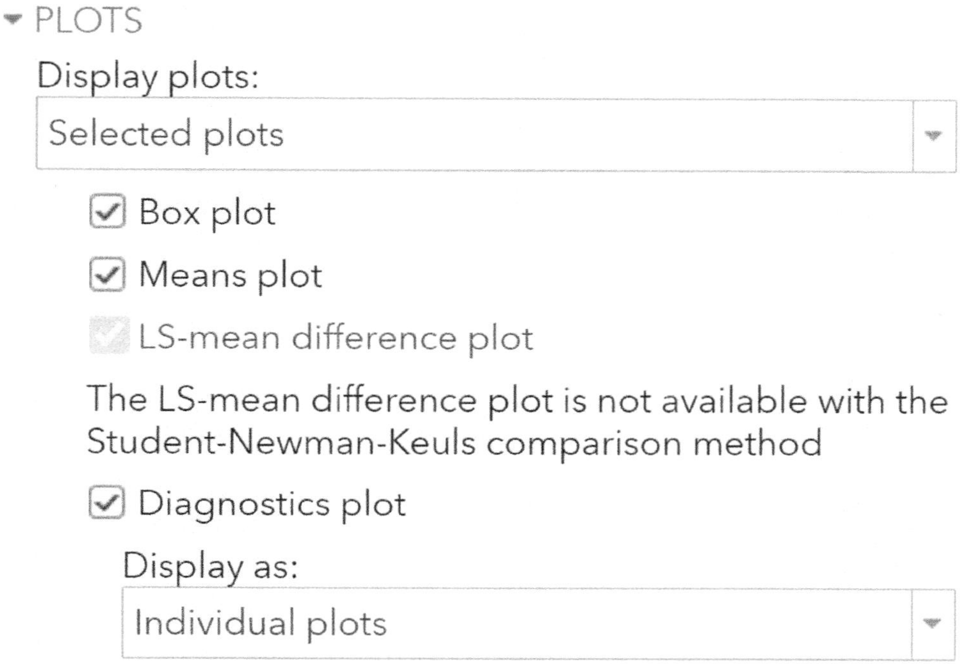

▾ PLOTS

Display plots:

Selected plots

☑ Box plot

☑ Means plot

☐ LS-mean difference plot

The LS-mean difference plot is not available with the Student-Newman-Keuls comparison method

☑ Diagnostics plot

Display as:

Individual plots

Figure 9.13: Requesting More Diagnostic Plots

The next several plots are intended to help you decide if the ANOVA assumptions were satisfied and to graphically show you information about the three means and the distribution of scores in each of the three groups.

> The figures shown below were selected from a larger set of plots produced by the one-way ANOVA task.

The plot shown in Figure 9.14 shows the residuals (the differences between the mean of each group and each individual score in that group). There are actually two residual plots produced by the one-way task. One (shown here) displays the residuals as actual scores (words-per-minute, in this example). Another residual plot (not shown) displays the residuals as *t* scores (the number of standard deviations above or below the mean of the group). Both plots look very similar. You also see the predicted values (means of each group) shown on the X axis.

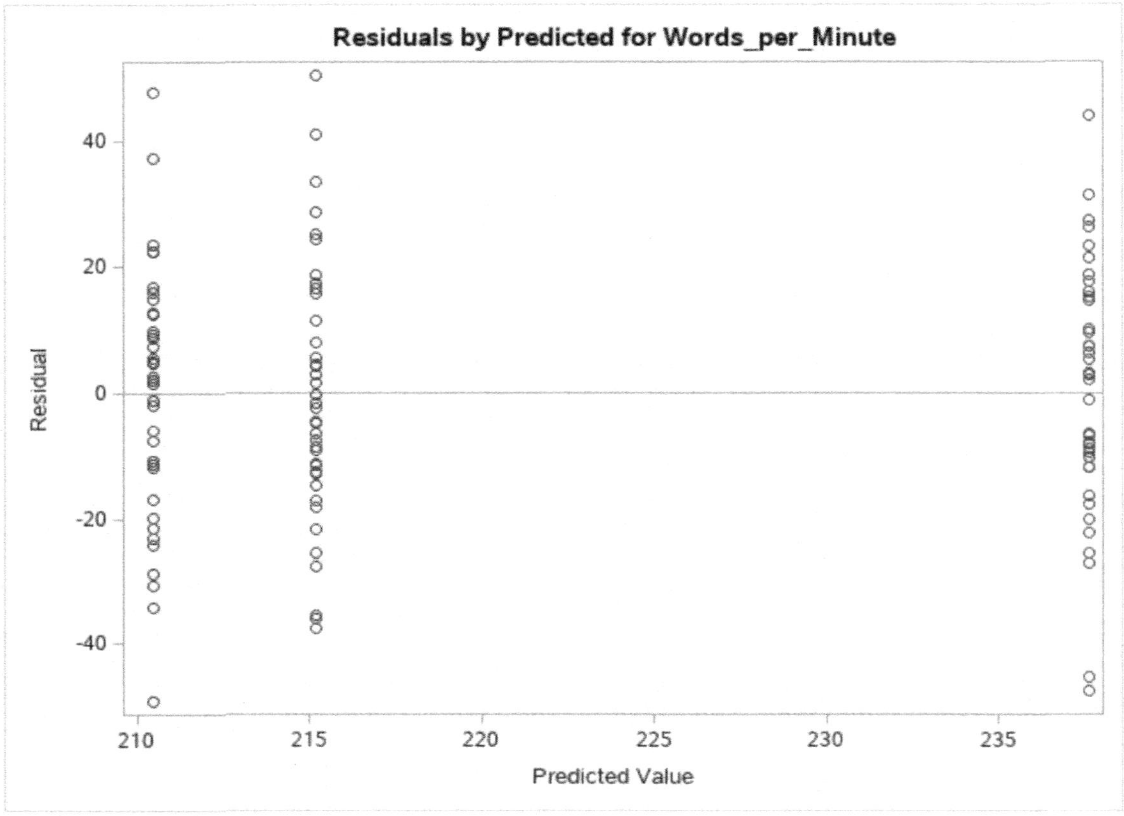

Figure 9.14: Residuals by Predicted Values

Notice that the residuals are spread out equally above and below the zero value for each of the three groups. This is another way to see that the variances in the three groups are not significantly different (as shown by the Levene Test).

One of the assumptions for running a one-way ANOVA is that the errors (the residuals are estimates of these errors) are normally distributed. You have seen Q-Q plots earlier in this book, so you remember that data values that are normally distributed appear as a straight line on a Q-Q plot. The plot shown in Figure 9.15 shows small deviations from a straight line, but not enough to invalidate the analysis.

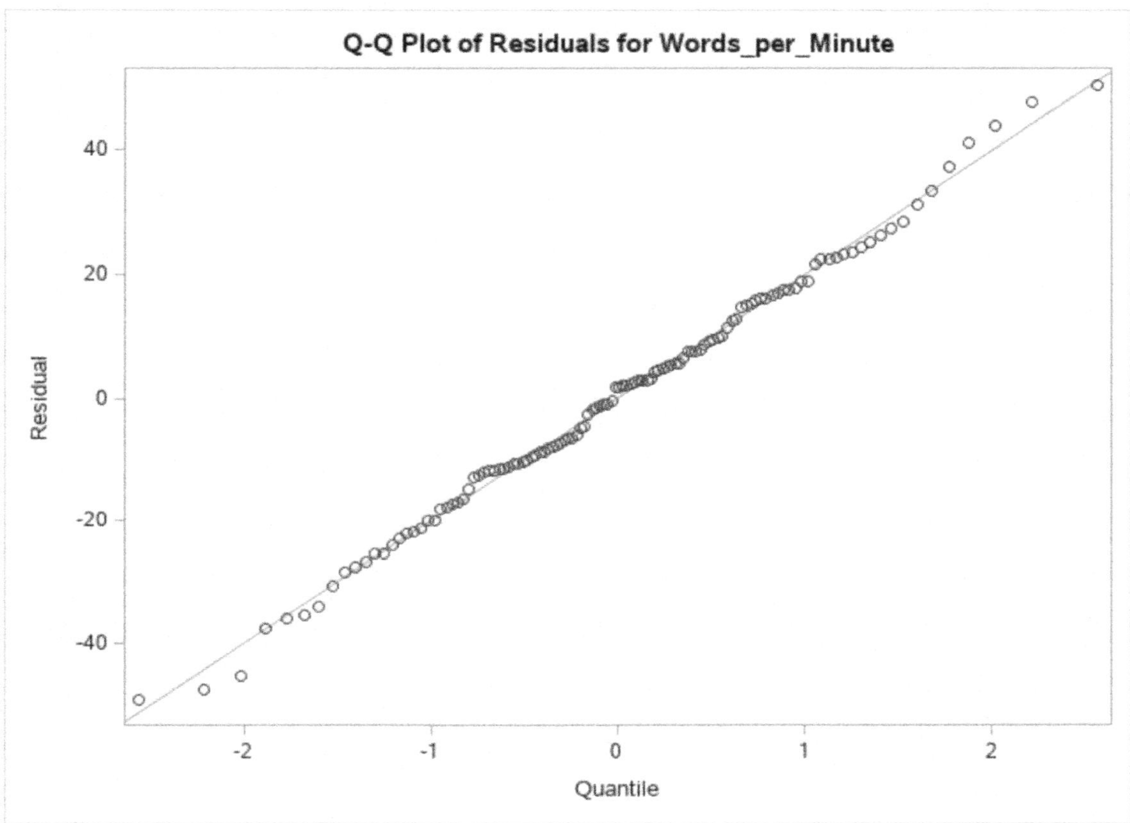

Figure 9.15: Q-Q Plot of the Residuals

Performing a Nonparametric One-Way Test

If you feel that the distribution assumptions are not satisfied by your data, another statistical task, Nonparametric One-Way ANOVA, provides a host of alternate tests. To demonstrate this, let's go back to the SASHELP data set called Fish and compare the weights of three species of fish.

Start out by selecting Nonparametric One-Way ANOVA found under the Linear Models in the Tasks list (Figure 9.16).

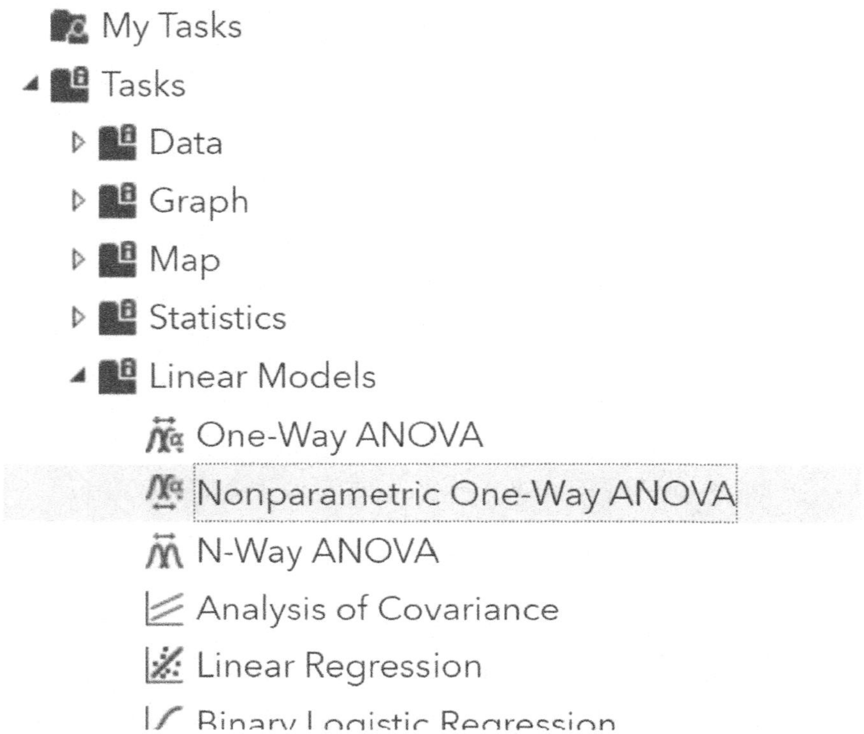

My Tasks
Tasks
Data
Graph
Map
Statistics
Linear Models
One-Way ANOVA
Nonparametric One-Way ANOVA
N-Way ANOVA
Analysis of Covariance
Linear Regression
Binary Logistic Regression

Figure 9.16: Select Nonparametric One-Way ANOVA from the Linear Models Tab

Next, you want to create a filter to select three species: Bream, Roach, and Pike. The expression that you need to write is shown in Figure 9.17.

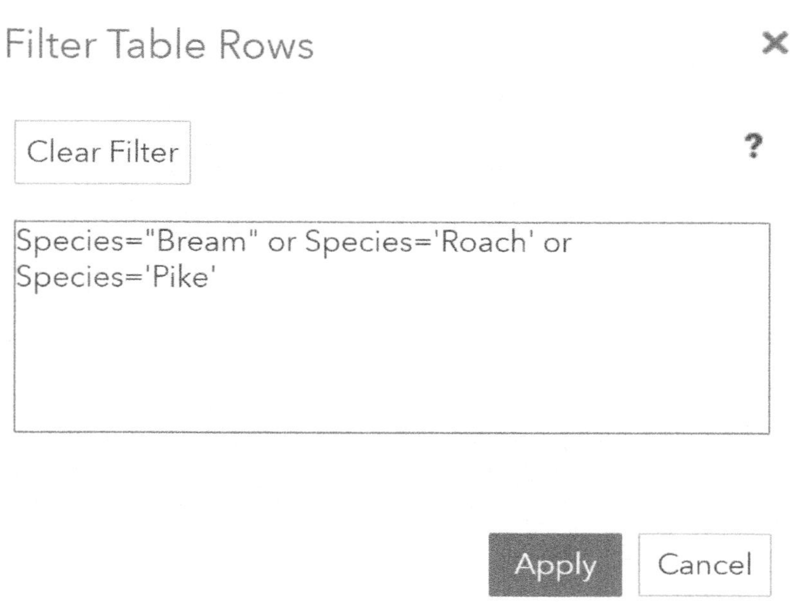

Figure 9.17: Creating a Filter for Three Species

The names of the three species need to be placed in either single or double quotation marks because Species is a character variable. Once you apply this filter, only the three species will be used in the analysis.

On the DATA Tab, select Weight as the Dependent variable and Species as the Classification variable (Figure 9.18).

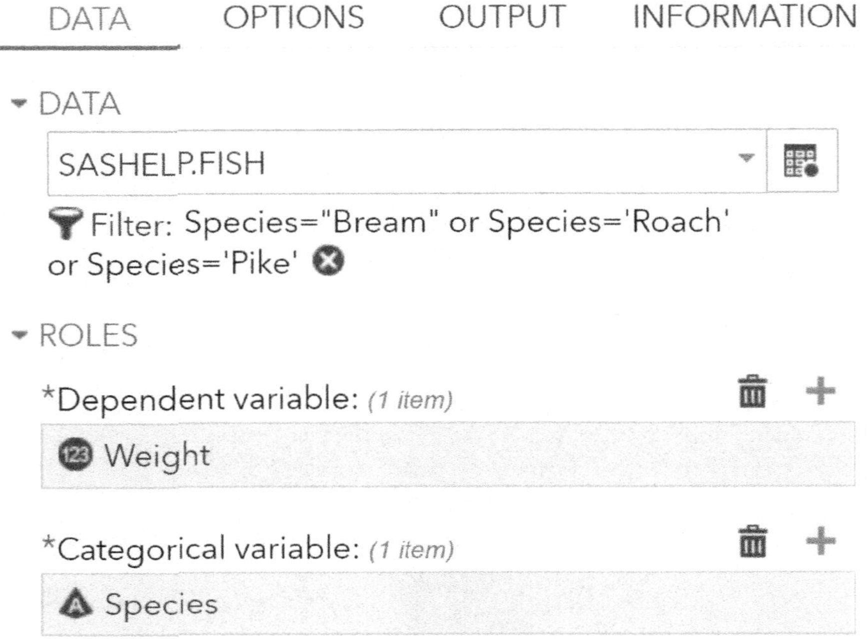

Figure 9.18: Identifying the Dependent and Classification Variables

Next, click the OPTIONS tab. For this example, you are using all the default values except for a request for Pairwise multiple comparison analysis (asymptotic only).

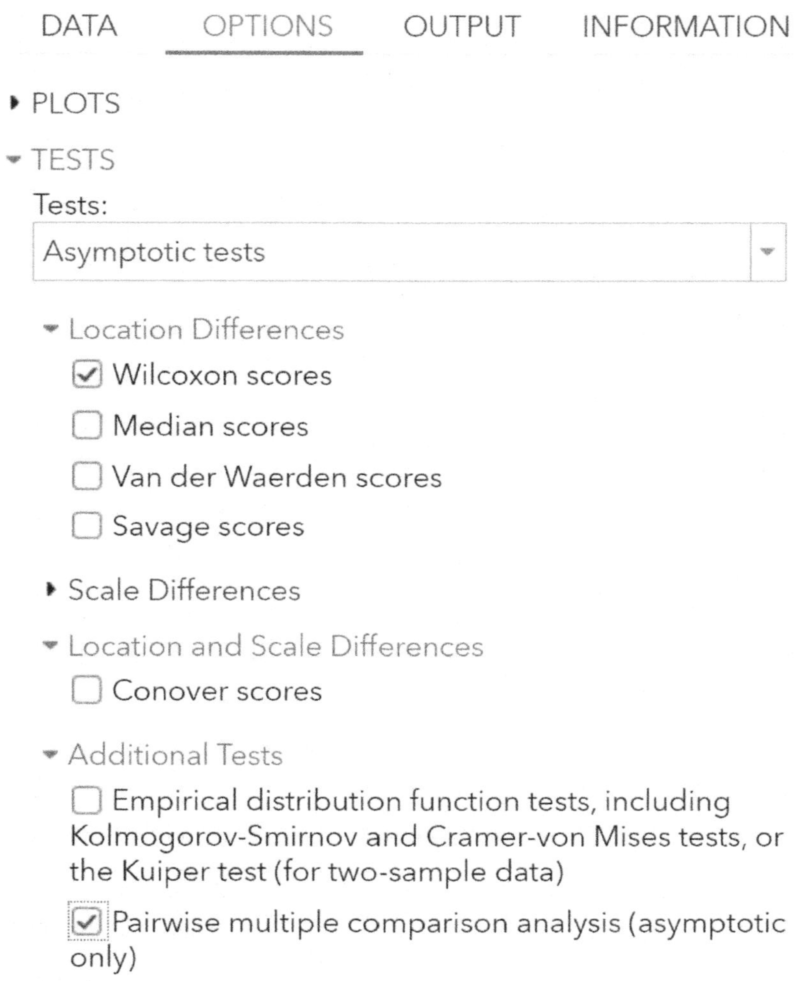

DATA OPTIONS OUTPUT INFORMATION

▸ PLOTS

▾ TESTS

Tests:

Asymptotic tests

▾ Location Differences

☑ Wilcoxon scores

☐ Median scores

☐ Van der Waerden scores

☐ Savage scores

▸ Scale Differences

▾ Location and Scale Differences

☐ Conover scores

▾ Additional Tests

☐ Empirical distribution function tests, including Kolmogorov-Smirnov and Cramer-von Mises tests, or the Kuiper test (for two-sample data)

☑ Pairwise multiple comparison analysis (asymptotic only)

Figure 9.19: Options for the Analysis

Now that you have entered your choices on the DATA and OPTIONS tabs, it's time to run the task. The first part of the output is shown in Figure 9.20.

Wilcoxon Scores (Rank Sums) for Variable Weight Classified by Variable Species					
Species	N	Sum of Scores	Expected Under H0	Std Dev Under H0	Mean Score
Bream	34	1580.00	1224.0	86.852273	46.470588
Roach	20	224.50	720.0	78.206158	11.225000
Pike	17	751.50	612.0	74.192876	44.205882
Average scores were used for ties.					

Kruskal-Wallis Test		
Chi-Square	DF	Pr > ChiSq
40.2791	2	<.0001

Figure 9.20: Results from the Wilcoxon Rank Test

To perform the Wilcoxon Rank Sum Test, all the scores are sorted in order from the lowest score to the highest score. You then assign a rank to each score: the lowest score is rank 1, the next highest score is rank 2, and so on. If several scores are equal, the Wilcoxon test assigns the mean rank to each of the values (don't worry about this detail for now). Along with the fish weights, you also know the species associated with each rank. There is a total of 71 fish weights (34 + 20 + 17), so the ranks range from 1 to 71. If the null hypothesis is that the fish are all about the same weight, you would expect the sum of ranks for each species to be about the same. You use this idea to form your null hypothesis. If one or more of the fish species has a very high or low sum of ranks, you might expect that there are differences in weight based on species.

The table above shows the sum of ranks for each of the fish species and the expected value if the null hypothesis is true. Notice that the sum of Scores for Roach (224.5) is much lower than the sums for Bream or Pike, making you suspect that Roach are typically lighter than either Bream or Pike. To decide whether you should reject the null hypothesis that all the three species have equal weights, you look at the *p*-value at the bottom of Figure 9.20. Because the *p*-value is shown as <.0001, you reject the null hypothesis and conclude that one or more pairs of means are significantly different. But, which pairs of fish species are different? We will answer that question in a minute.

The next part of the output shows box plots for each fish species. Box plots are appropriate for this display because you are conducting a nonparametric test.

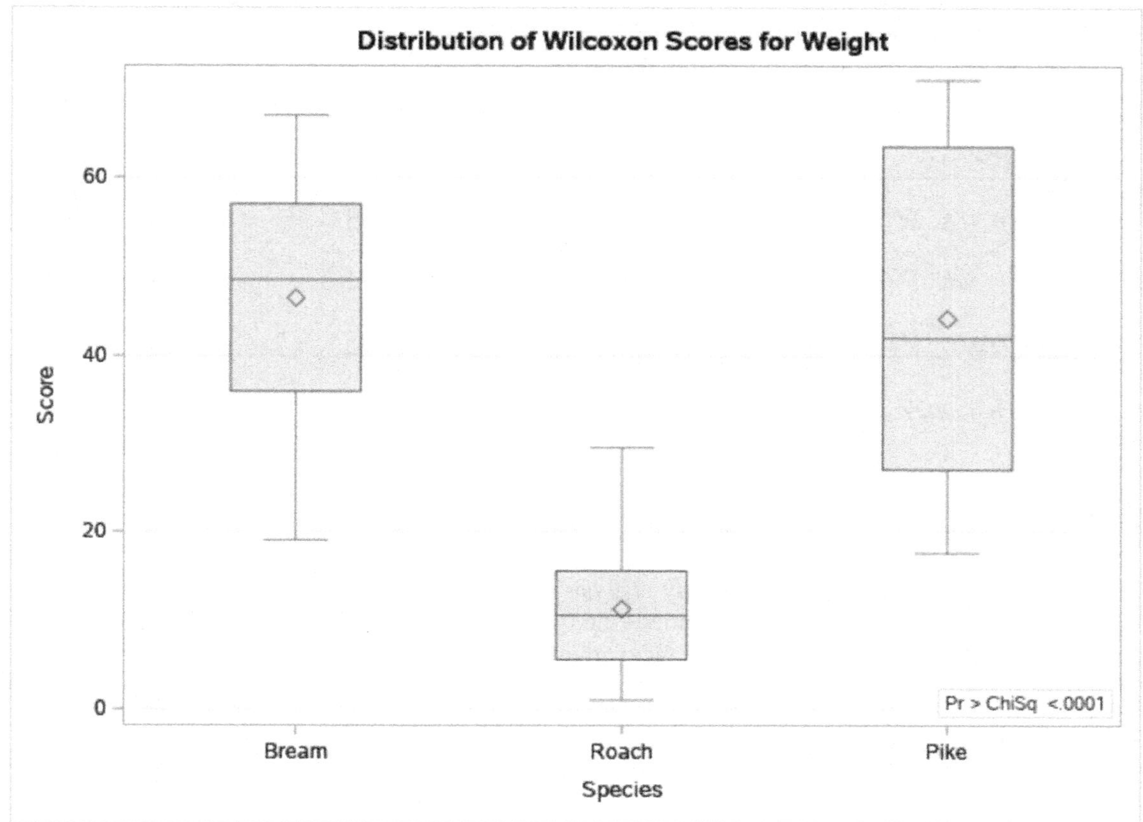

Figure 9.21: Box Plots for Each Species

This plot shows that Roach seem to be lighter than either Bream or Pike. Just as you did a multiple comparison test in the ANOVA test (Tukey or SNK), there is an equivalent multiple comparison nonparametric test. Figure 9.22 shows significant differences between Bream versus Roach and Roach versus Pike. Bream and Pike are not statistically different (p = .8900).

Pairwise Two-Sided Multiple Comparison Analysis			
Dwass, Steel, Critchlow-Fligner Method			
Variable: Weight			
Species	Wilcoxon Z	DSCF Value	Pr > DSCF
Bream vs. Roach	5.9671	8.4388	<.0001
Bream vs. Pike	0.4599	0.6504	0.8900
Roach vs. Pike	-4.9544	7.0066	<.0001

Figure 9.22: Pairwise Comparisons

Conclusion

You have seen how to conduct a one-way analysis of variance as well as a Wilcoxon nonparametric test. You have also seen ways to determine if the two assumptions for a one-way ANOVA (normally distributed data and homogeneity of variance) are met.

Chapter 9 Exercises

1. List the first 10 observations from the High_School data set found in the STATS library (this was created when you ran the Create_Dataset.sas program in the download package). Conduct a one-way ANOVA comparing the variable Vocab_Score (a measure of vocabulary skill) by Grade (Freshman, Sophomore, Junior, and Senior). Be sure to run a Tukey multiple comparison test to determine which grades are different from each other (or none).
2. Repeat exercise 1, except this time, use the variable English_Grade as the dependent variable.
3. The data set Salary_Formatted in the STATS library contains variables Gender, Age_Group, and Weekly_Salary. First, run the short program below to create a new variable (Gender_Age) that creates four combinations of the two variables Gender and Age_Group. (Hint: click the new options icon and request a New Program.)

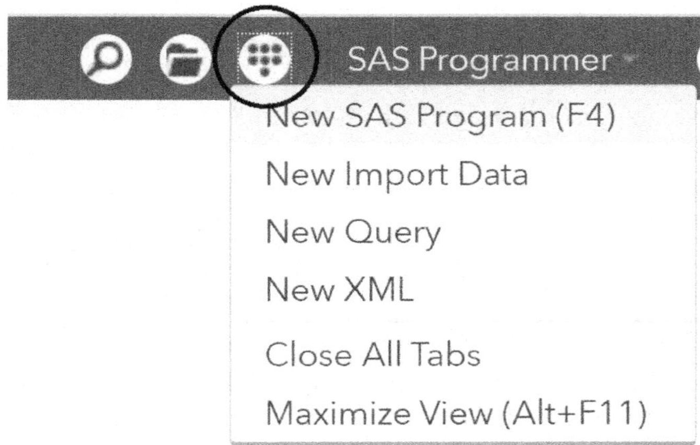


```
data Temp;
   set Stats.Salary_Formatted;
   length Gender_Age $ 6;
   Gender_Age = Cats(Gender, Age_Group);
run;
```

The first 10 observations in the Temp data set should look like this:

Obs	Gender	Age_Group	Education	Weekly_Salary	Salary	Gender_Age
1	F	20-24	<HS	$443	0	F20-24
2	F	20-24	BA+	$1,574	1	F20-24
3	F	45-54	<HS	$1,089	0	F45-54
4	F	45-54	BA+	$1,861	1	F45-54
5	M	20-24	<HS	$939	0	M20-24
6	M	20-24	<HS	$872	0	M20-24
7	M	20-24	<HS	$819	0	M20-24
8	M	20-24	<HS	$873	0	M20-24
9	M	20-24	<HS	$970	0	M20-24
10	M	20-24	<HS	$680	0	M20-24

Using the Temp data set, run a one-way analysis using Weekly_Salary as the dependent variable and Gender_Age as the categorical variable.

4. Using the data set Fish in the SASHELP library, compare the Width (not the Weight) of three species of fish; Perch, Roach, and Pike. Do this using both parametric and nonparametric methods. You will need to create a filter that

reads:

```
Species='Perch' or Species='Roach' or Species='Pike'
```

5. Using the data set Heart in the SASHELP library, run a nonparametric ANOVA using Cholesterol as your dependent variable and DeathCause as your classification variable. Include an option for multiple comparisons. Which, if any, of the causes of death had significant differences in cholesterol?
6. Using the data set Cars in the SASHELP library, compare the Horsepower for each Type of car. Use a filter (Type ne 'Hybrid') to eliminate hybrids because there are so few of them.
7. Using the data set Cars in the SASHELP library, compare the Weight of four-, six-, and eight-cylinder cars. Use a filter to restrict the variable Cylinders to values of 4, 6, or 8. Hint: the filter expression should read: Cylinders = 4 or Cylinders=6 or Cylinders=8. An interesting alternative is: Cylinders IN (4,6,8).

10.

N-Way ANOVA

Introduction

You can construct ANOVA models with more than one independent variable. One of the most popular models is called a **factorial model**. In a factorial model, you compute variances for each independent variable as well as interaction terms.

When you run an analysis of variance with more than one independent variable, you can look at the role of each variable after controlling for any variability attributed to the other variables. This gives you a better chance of understanding the unique role of each variable in the model. Things get a bit complicated if the effect of one variable interacts with the effect of another variable or variables. This is known as an interaction, and you will see later in this chapter how to detect and analyze models where significant interactions are present.

A data set called Exercise was created when you ran the program Create_Datasets.sas (located in the download from SAS from the book's web page). This data set, located in the STATS library, contains variables Gender, Training (10, 20, or 30 minutes per session), and Dose (placebo or 100 mg of a drug thought to increase strength), along with three outcome variables Pushups (the number of push-ups a person can do), Strength, and Flexibility.

For this first example, you want to see the effect of Gender and Training on the variable Pushups. Later in the chapter, you will test the effects of Gender and Dose (placebo or active drug) on the variable Strength (on a scale where higher numbers indicate more strength). A third dependent variable (Flexibility) (on a scale where higher numbers represent more flexibility) will be used in an end-of-chapter exercise.

The Linear Models task N-Way ANOVA helps you analyze factorial models as well as more advanced models that involve crossing and nesting.

Performing a Two-Way Analysis of Variance

You start an N-Way ANOVA by clicking Tasks ⇒ Linear Models ⇒ N-Way ANOVA. (See Figure 10.1 below.)

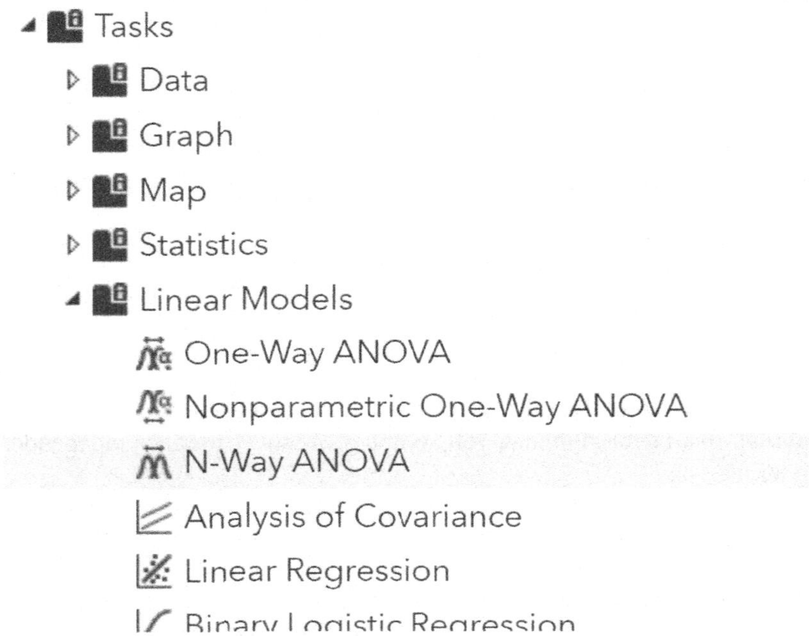

Tasks
Data
Graph
Map
Statistics
Linear Models
One-Way ANOVA
Nonparametric One-Way ANOVA
N-Way ANOVA
Analysis of Covariance
Linear Regression
Binary Logistic Regression

Figure 10.1: Select N-Way ANOVA

Double-click N-Way ANOVA to begin:

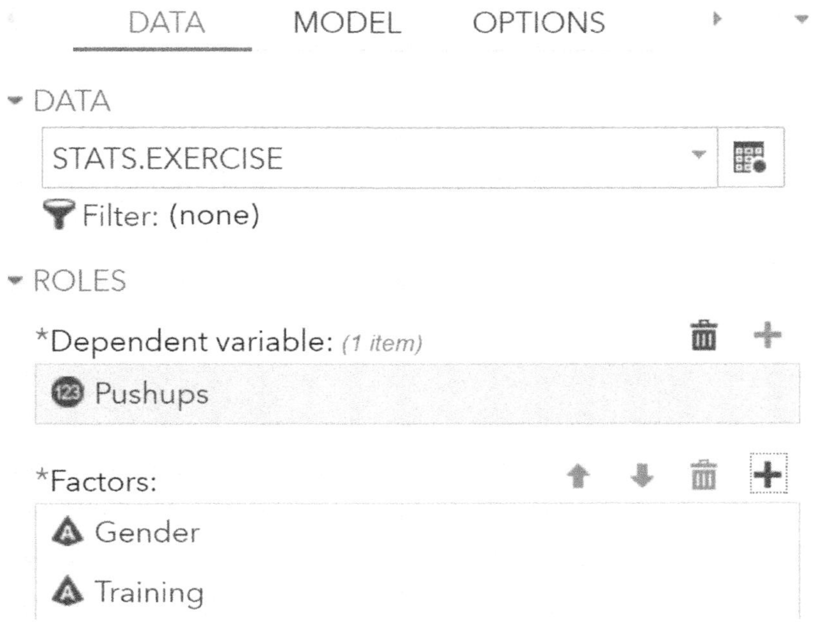

DATA MODEL OPTIONS

DATA

STATS.EXERCISE

Filter: (none)

ROLES

*Dependent variable: (1 item)

Pushups

*Factors:

Gender

Training

Figure 10.2: Select Your Dependent Variable and Factors

You selected the Exercise data set in the STATS library, Pushups as the Dependent variable, and Gender and Training as Factors (also called independent variables).

Next, click the MODEL tab to select a model. The MODEL tab is shown in Figure 10.3. You must first click the Edit button (circled in the figure below).

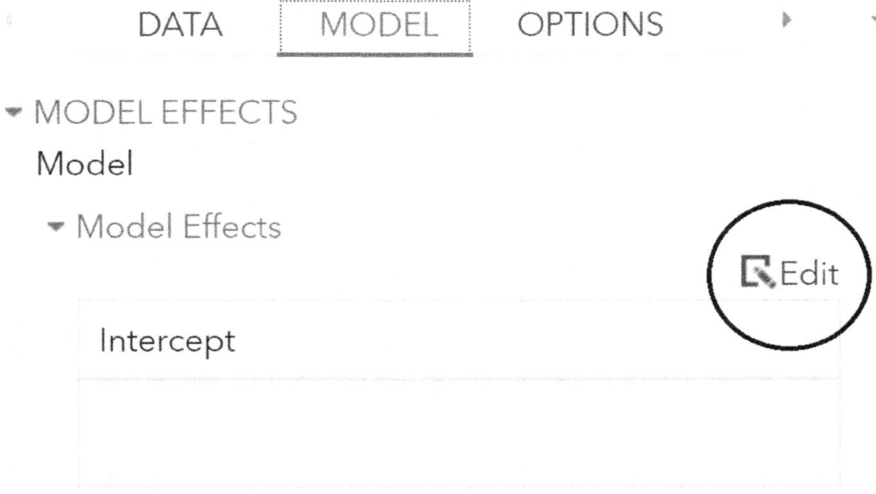

Figure 10.3: Click Edit in the MODEL Window

This brings you to the model building screen shown in Figure 10.4.

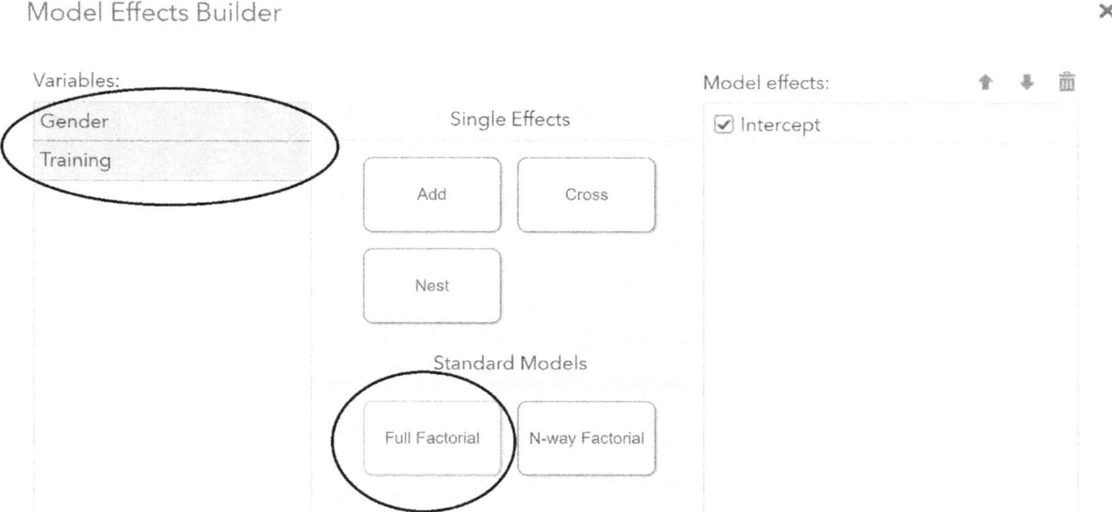

Figure 10.4: Enter Your Independent and Full Factorial

You can select the variables one at a time or hold down the ctrl key and select them all at once. Once you have selected the independent variables, click the Add box to add them to the model, then click the box labeled Full Factorial. Your model is then shown on the right side of the screen as shown in Figure 10.5.

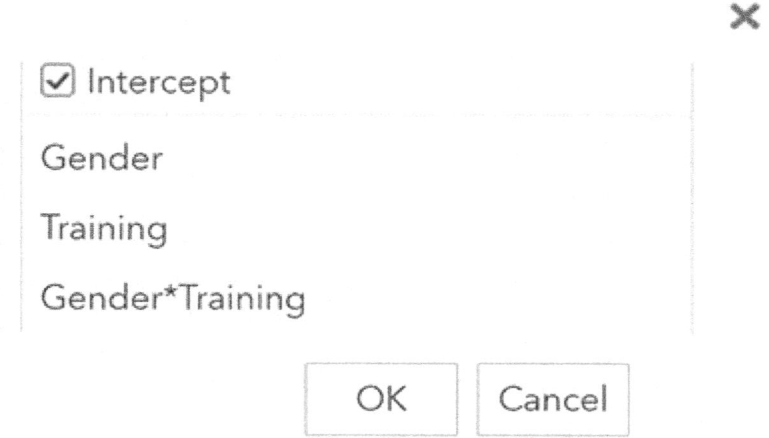

Figure 10.5: View Your Model and Click OK at the Bottom of the Window

The model shows your two variables, Gender and Training. The term Gender*Training is standard notation for an interaction between these two variables. It is important to scroll down to the bottom of this screen and click OK. The reason this is emphasized is that this author did not see the OK button the first time he ran this model and was confused when nothing happened.

The last step is to click the OPTIONS tab. It is here that you can make decisions about what analyses to run and the type of plots you want to see. To specify what statistics to display, choose 'Default and additional statistics' from the pull-down list. Be sure the box to perform multiple comparisons is checked (that is usually the default). The default method is Tukey (a popular choice), and the significance level is set at .05. Finally, under Sums of Squares, deselect Sequential sum of squares (Type 1) and be sure the Adjusted sum of squares (Type 3) is checked. Type 1 sum of squares is a sequential sum of squares that adds each variable one at a time in the order specified by your model—the Type 3 sum of squares shows the effects of each variable after controlling for the remaining variables in the model. This is the usual choice for an analysis of variance.

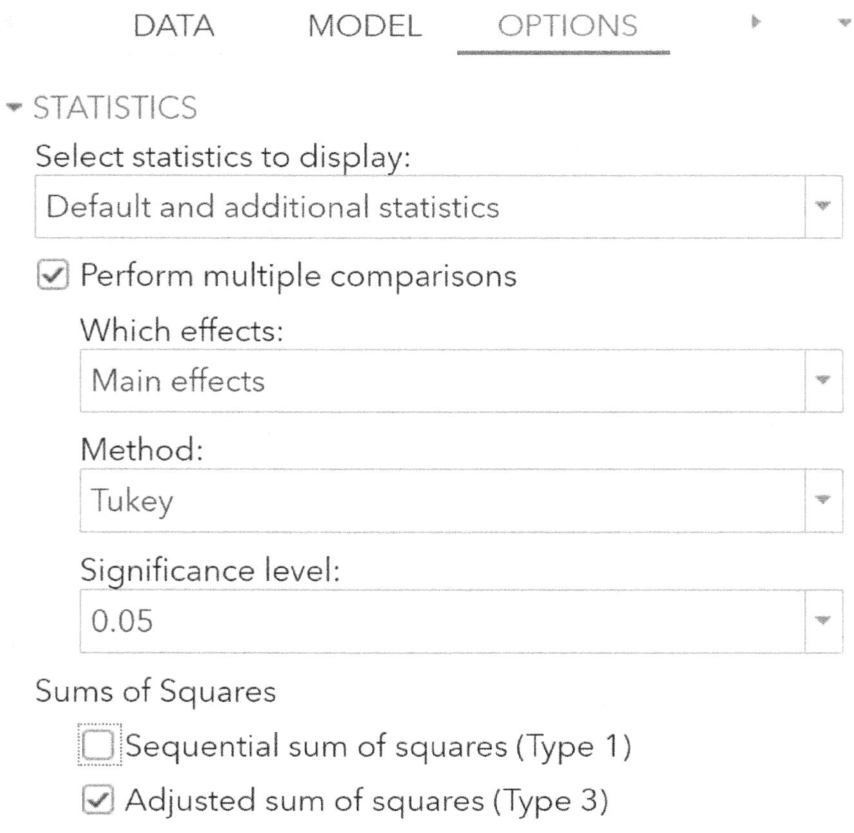

Figure 10.6: Make Selections on the OPTIONS Tab

At the bottom of the screen, you can select the plots that you want to see and how they are presented. (See Figure 10.7.)

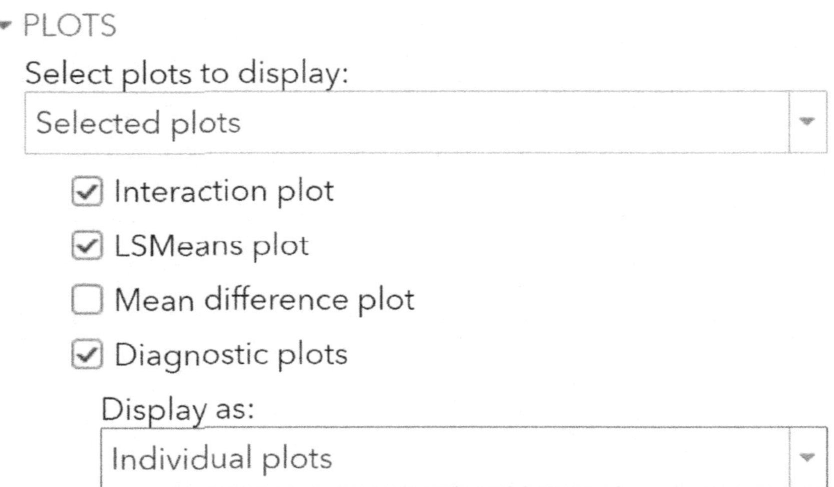

Figure 10.7: Select Plots

Notice that Selected plots has been chosen along with an interaction plot, an LSMeans plot, and Diagnostic plots. Notice also that the Diagnostic plots are to be displayed as individual plots. The other choice is a panel of plots where all the diagnostic plots are grouped together in a single panel.

Click the Run icon to obtain the output below.

Class Level Information		
Class	**Levels**	**Values**
Gender	2	F M
Training	3	10 20 30

Number of Observations Read	240
Number of Observations Used	240

Figure 10.8: Class Level Information

The first part of output is similar to the Class Level Information that you saw in the One-Way ANOVA described in the previous chapter. Once again, this is important information! Be sure you see the correct values for each level and that you have the number of levels that you expect. Any data error in either variable (Gender or Training) will result in additional levels and make your analysis invalid. Check the Number of Observations Read and the Number of Observations Used to see how many missing values you had in your data. In this example, there were no missing values because the two numbers are equal.

Next, move on to the ANOVA Table.

Dependent Variable: Pushups

Source	DF	Sum of Squares	Mean Square	F Value	Pr > F
Model	5	20207.72083	4041.54417	150.32	<.0001
Error	234	6291.57500	26.88707		
Corrected Total	239	26499.29583			

R-Square	Coeff Var	Root MSE	Pushups Mean
0.762576	11.66650	5.185275	44.44583

Source	DF	Type III SS	Mean Square	F Value	Pr > F
Gender	1	6417.00417	6417.00417	238.67	<.0001
Training	2	13764.30833	6882.15417	255.97	<.0001
Gender*Training	2	26.40833	13.20417	0.49	0.6126

Figure 10.9: ANOVA Table

The Mean Square for the model is 4041.54417, and the Mean Square for the Error is 26.88707, giving you an F Value of 150.32 (4041.54417 divided by 26.88707). Unless you have memorized the F-distribution for all possible degrees of freedom, you can see that the entire model is highly significant ($p < .0001$).

Notice that the mean number of push-ups is 44.44583 (pretty impressive group of subjects). Next, look at the probabilities for the Gender and Training as well as the interaction term (Gender*Training). Take a big sigh of relief that the interaction is nowhere near significant ($p = .6126$), telling you that you can confidently interpret the main effects. You can say that both Gender and Training are highly significant in determining how many push-ups a person can do.

Reviewing the Diagnostic Plots

You are not finished. You need to determine if two assumptions for an analysis of variance were reasonably satisfied. First, you want to inspect the distribution of residuals. These are the differences between the mean of each combination of Gender and Training and each individual's number of push-ups. SAS Studio produces two residual plots: one shows the residuals as actual values (number of push-ups), and the other shows these values transformed into *t*-scores. Because they look almost identical, the Studentized (*t*-score) residuals are shown in Figure 10.10. The distributions look uniform with only a few points outside, plus or minus 2 standard deviations, and the variances do not look drastically different.

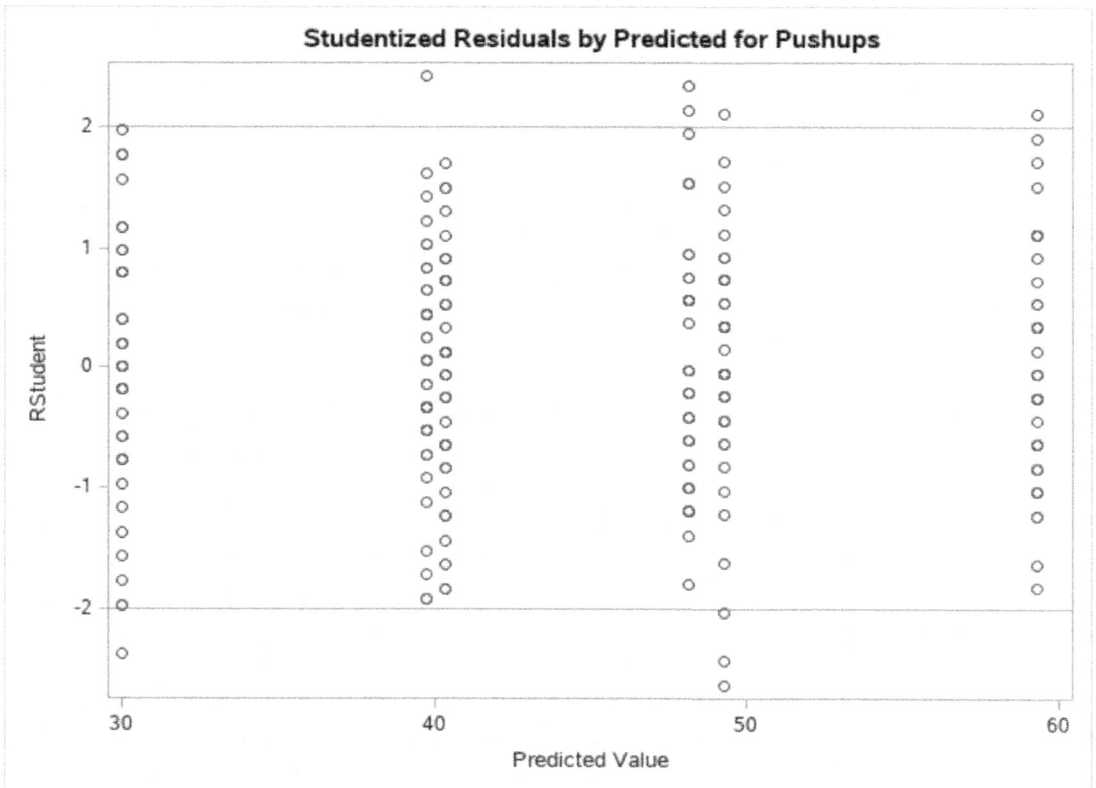

Figure 10.10: Residual Plot

The distribution of residuals shown in Figure 10.11 looks very close to normal.

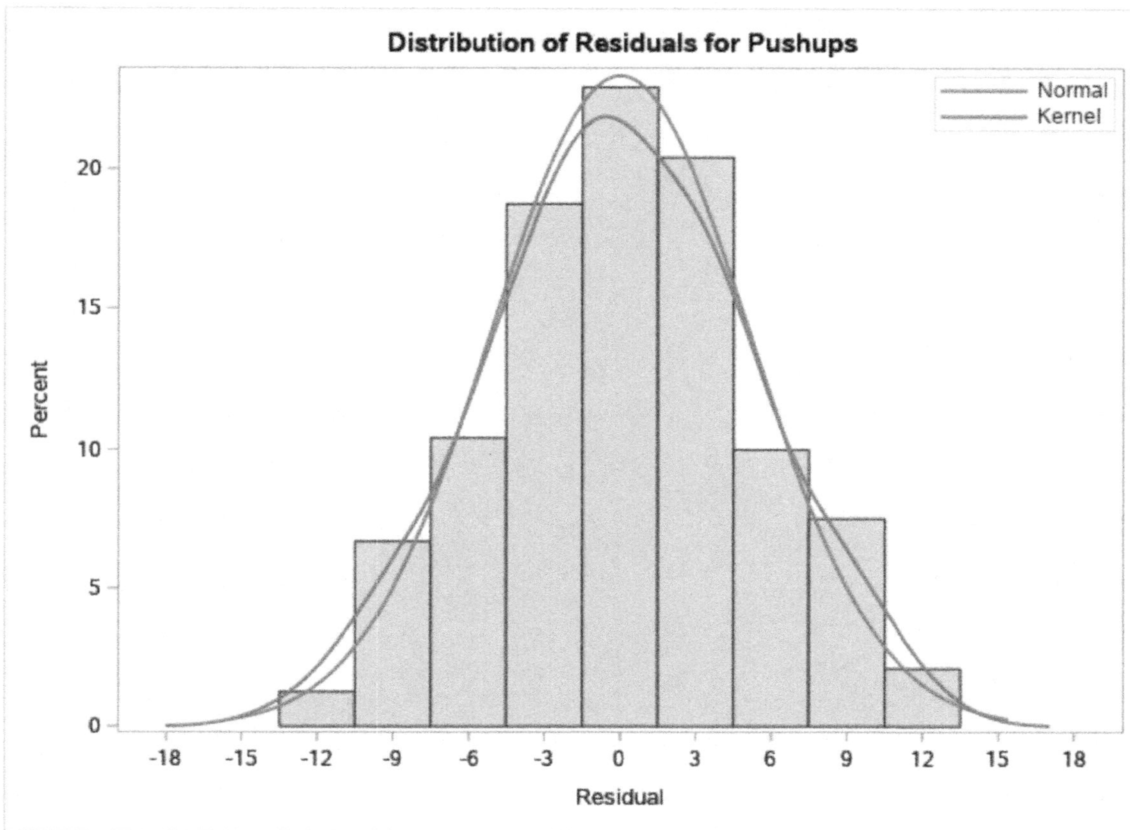

Figure 10.11: Distribution of Residuals

An interaction plot is displayed in Figure 10.12. You see the effect of training for males and females. Each of the three lines in the plot are parallel, demonstrating that the training effect is independent of gender. You already knew that from the high *p*-value for the interaction term in the model.

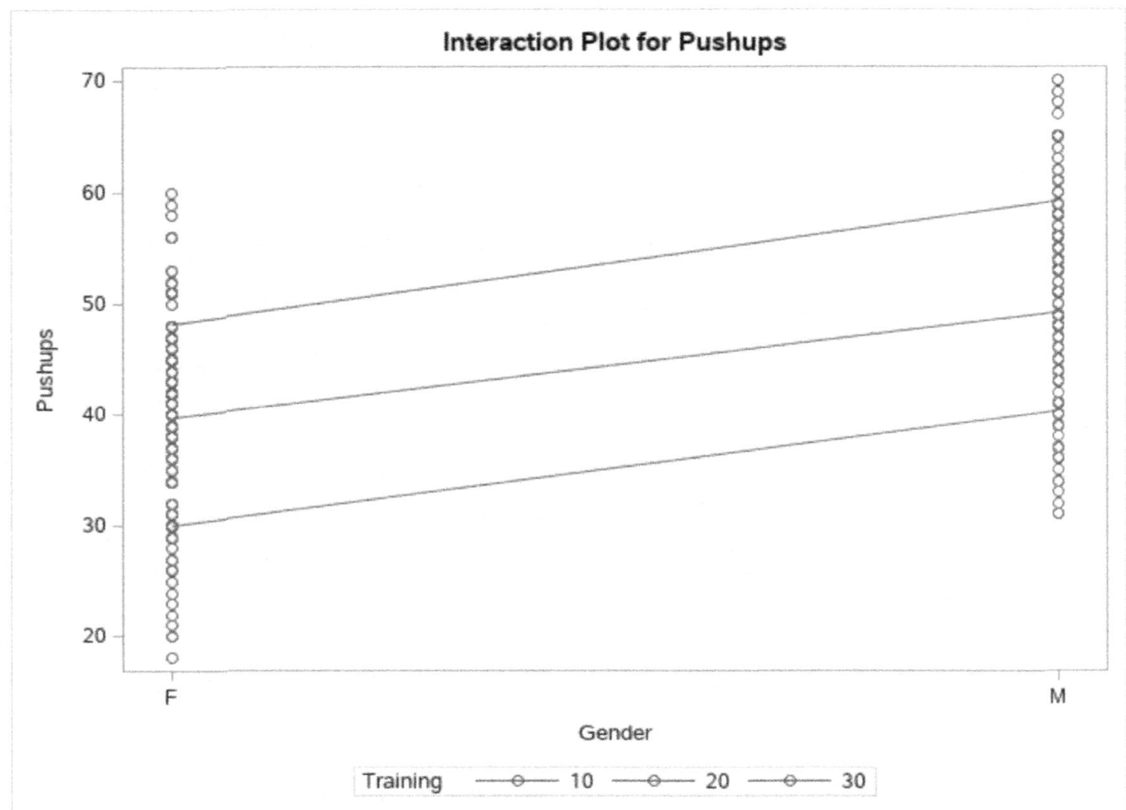

Figure 10.12: Interaction Plot

LSmeans (least square means) are computed from the linear model. If there are no missing values in your data (the case here), these means are exactly the same means that you would compute by adding up the number of push-ups for each combination of gender and training level and dividing by the number of subjects in that group. If there are any missing values for any observations, the LSMeans will produce slightly different values.

Least Squares Means
Adjustment for Multiple Comparisons: Tukey

Gender	Pushups LSMEAN	H0:LSMean1=LSMean2
		Pr > |t|
F	39.2750000	<.0001
M	49.6166667	

Figure 10.13: Least Squares Means

You see in Figure 10.13 that the means for each level of Gender (males did about 10 more than women). In Figure 10.14, you can see that the more training people receive, the more push-ups they can do.

Least Squares Means
Adjustment for Multiple Comparisons: Tukey

Training	Pushups LSMEAN	LSMEAN Number
10	35.1500000	1
20	44.4875000	2
30	53.7000000	3

Least Squares Means for effect Training Pr > |t| for H0: LSMean(i)=LSMean(j) Dependent Variable: Pushups			
i/j	1	2	3
1		<.0001	<.0001
2	<.0001		<.0001
3	<.0001	<.0001	

Figure 10.14: Tukey Multiple Comparisons

Because there are two levels for Gender, and Gender was highly significant, you know that the comparison of males to females is significant. There are three levels of training. In order to decide which levels are different, you can look at the probabilities shown in the bottom of Figure 10.14. Each level of training is assigned an LSMEAN Number. The matrix at the very bottom of the figure shows you the *p*-value for each comparison. Training level 1 is significantly different from training level 2 and training level 2 is significantly different from training level 3 (all *p*-values <.0001).

Interpreting Models with Significant Interactions

For this section, we are going to use the two independent variables, Gender and Dose, to see how they affect a person's strength. Begin as before. Select the data set, independent variable, and factors on the DATA tab (Figure 10.15).

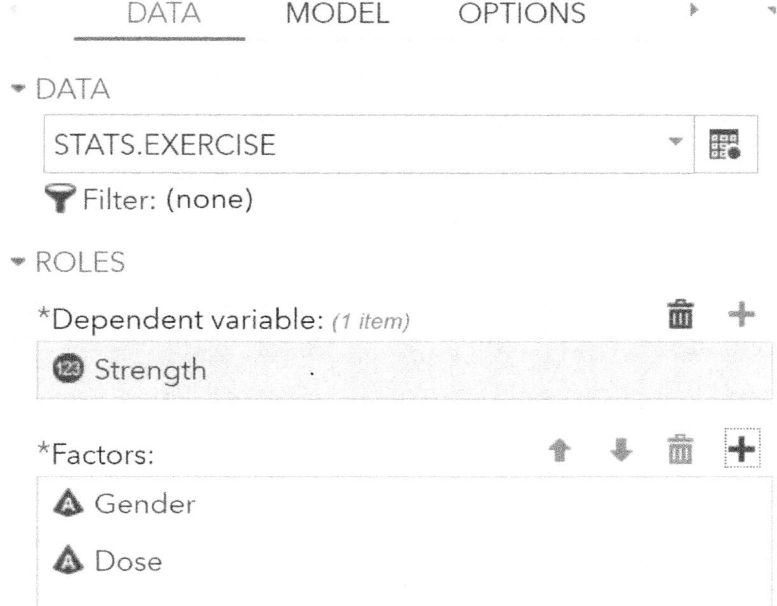

Figure 10.15: DATA Tab for Two-Way ANOVA

Use the MODEL tab the same way you did for the previous example. Next, click the OPTIONS tab and request multiple comparisons and Type 3 sums of squares (Figure 10.16).

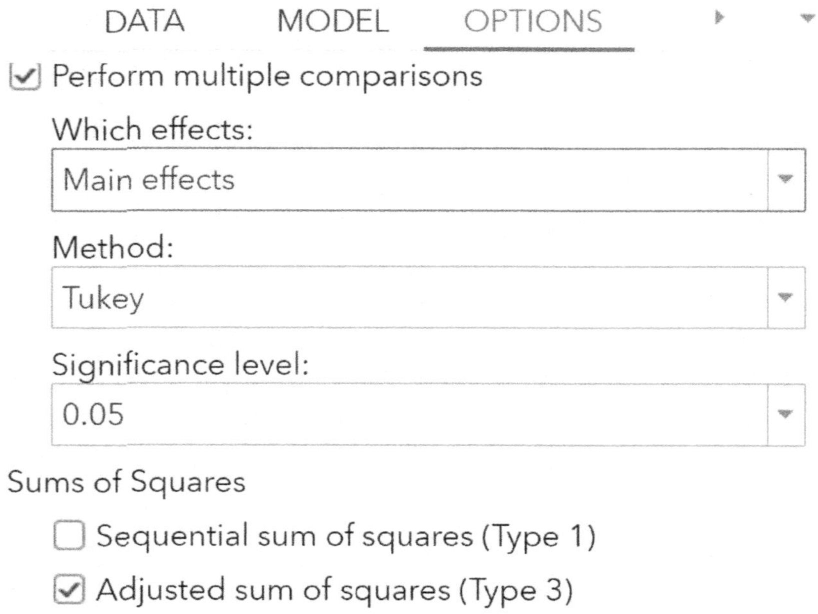

DATA MODEL OPTIONS

☑ Perform multiple comparisons

Which effects:

Main effects

Method:

Tukey

Significance level:

0.05

Sums of Squares

☐ Sequential sum of squares (Type 1)

☑ Adjusted sum of squares (Type 3)

Figure 10.16: Model Options

Click the Run icon to run the program. The first portion of the output is shown in Figure 10.17.

Class Level Information		
Class	**Levels**	**Values**
Gender	2	F M
Dose	2	100 mg Placebo

Number of Observations Read	240
Number of Observations Used	240

Figure 10.17: Class Level Information

You see that the values for Gender and Dose are correct and that there are no missing values (the number of observations read and used are the same).

Before you get all excited about the very small *p*-values in the ANOVA table shown in Figure 10.18, you should focus on the significant interaction at the bottom of the figure.

Dependent Variable: Strength

Source	DF	Sum of Squares	Mean Square	F Value	Pr > F
Model	3	185416.879	61805.626	5.90	0.0007
Error	236	2470584.283	10468.577		
Corrected Total	239	2656001.163			

R-Square	Coeff Var	Root MSE	Strength Mean
0.069811	82.57122	102.3161	123.9125

Source	DF	Type III SS	Mean Square	F Value	Pr > F
Gender	1	138288.0042	138288.0042	13.21	0.0003
Dose	1	2001.0375	2001.0375	0.19	0.6624
Gender*Dose	1	45127.8375	45127.8375	4.31	0.0390

Figure 10.18: Two-Way ANOVA Table

Once you have a significant interaction, it is more difficult to interpret the main effects. In this example, the difficulty resulting from the Gender by Dose interaction is that the influence of Dose is not the same for males and females. You will see shortly that there are steps that you can take to untangle the main effects. However, let's look at an interaction plot to see what is going on.

Go back to the OPTIONS tab and request an interaction plot (and a few other options) as shown in Figure 10.19.

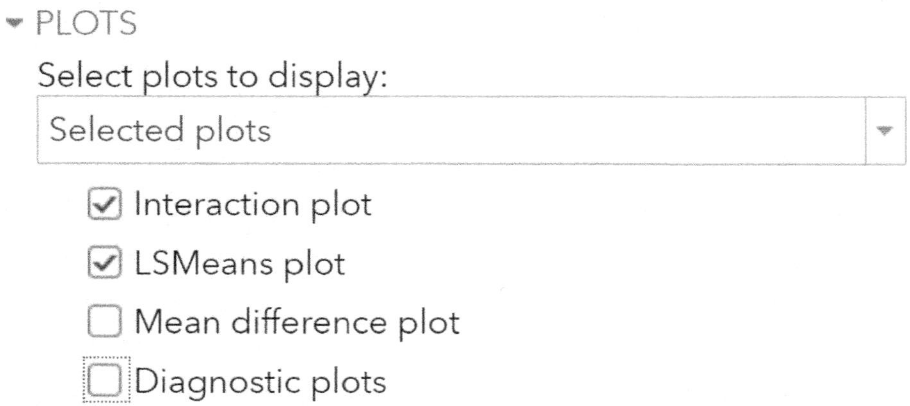

PLOTS

Select plots to display:

Selected plots

- [x] Interaction plot
- [x] LSMeans plot
- [ ] Mean difference plot
- [ ] Diagnostic plots

Figure 10.19: Requesting an Interaction Plot

Figure 10.20 shows the data points and means for the variable Strength for each of the two doses.

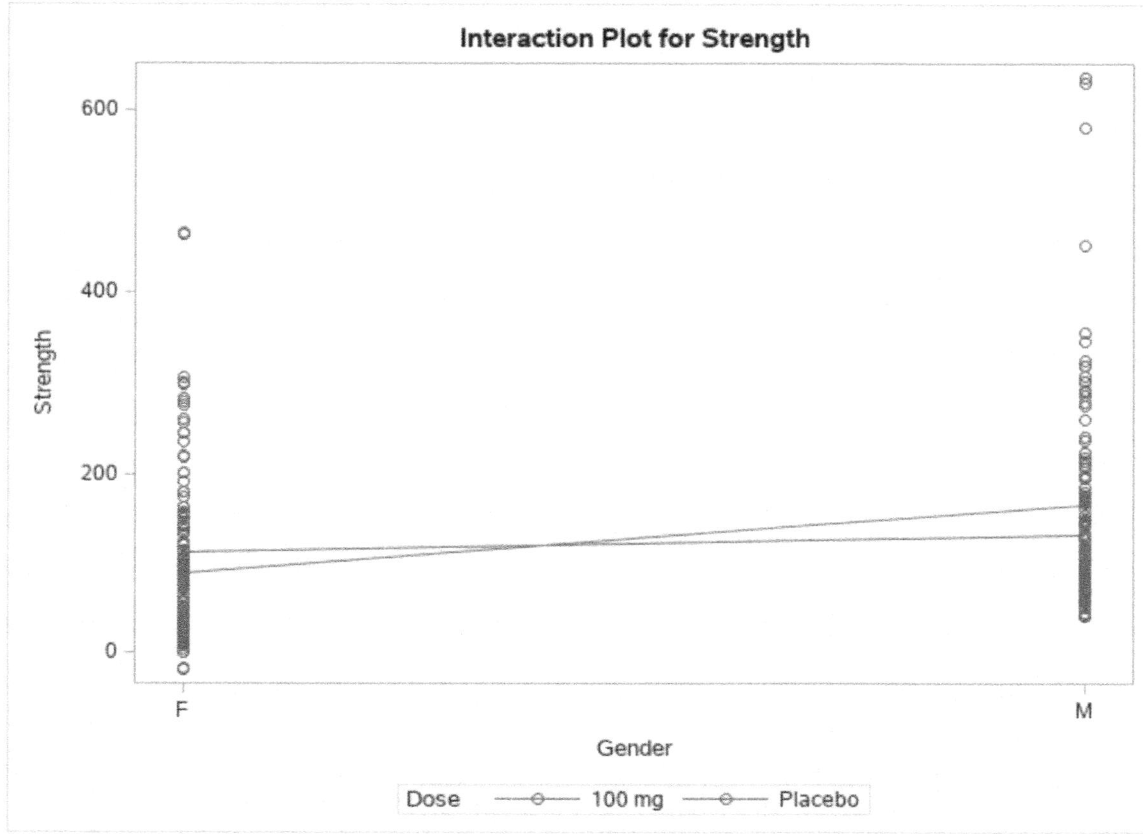

Figure 10.20: Interaction Plot

It seems that the active drug increases strength for males but decreases strength for females. Interaction plots for variables that show a significant interaction always show lines that are not parallel. In this case, they even cross. It's back to the drawing board.

Investigating the Interaction

Because of the significant interaction between Gender and Dose, you can go back to the STATISTICS tab and instead of the default action in the drop-down list (Default Statistics), select Default and Additional Statistics. (See Figure 10.21.)

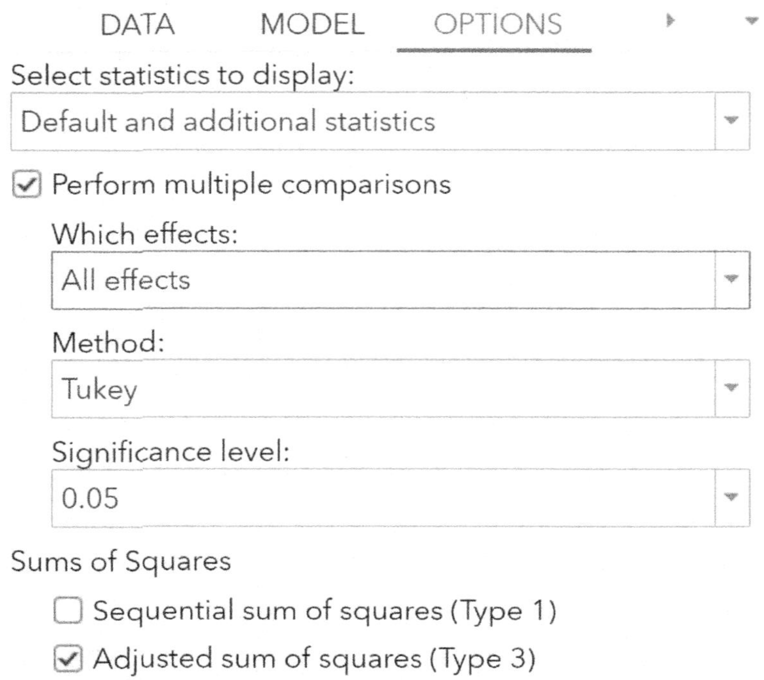

DATA MODEL OPTIONS

Select statistics to display:

Default and additional statistics

☑ Perform multiple comparisons

Which effects:

All effects

Method:

Tukey

Significance level:

0.05

Sums of Squares

☐ Sequential sum of squares (Type 1)

☑ Adjusted sum of squares (Type 3)

Figure 10.21: Requesting Multiple Comparisons Including Interactions

Select All effects from the section on multiple comparisons. By making this selection, the output will now show comparisons for each combination of Gender and Dose as shown in the figure below.

Least Squares Means
Adjustment for Multiple Comparisons: Tukey

Gender	Dose	Strength LSMEAN	LSMEAN Number
F	100 mg	89.083333	1
F	Placebo	110.733333	2
M	100 mg	164.516667	3
M	Placebo	131.316667	4

Least Squares Means for effect Gender*Dose
Pr > |t| for H0: LSMean(i)=LSMean(j)

Dependent Variable: Strength

i/j	1	2	3	4
1		0.6533	0.0004	0.1103
2	0.6533		0.0225	0.6887
3	0.0004	0.0225		0.2869
4	0.1103	0.6887	0.2869	

Figure 10.22: Means for Each Gender and Dose

This portion of the output shows you the adjusted means for every combination of Dose and Gender. Notice that Females have a lower mean for the 100 mg dose compared to placebo while Males had a higher mean for the 100mg dose compared to the placebo. Using the Tukey multiple comparison test, you can also see differences between every pair of values. For example, comparing the mean for females on the active drug (LSMean = 1) to females on placebo (LSMeans = 2), you see a *p*-value of .6533 (that is, not significant). A similar comparison for males (LSMeans = 3 and 4) is also not significant (p = .2869). Comparing males to females on the 100 mg dose (LSMeans 1 and 3), you see a highly significant difference (p = .0004).

Conclusion

This chapter has demonstrated simple two-way models, one without interactions and one with a significant interaction. Much more complex models with nested and/or crossed factors can be analyzed with the N-Way ANOVA task.

Chapter 10 Exercises

1. Using the data set High_School in the STATS library, test the effect of Gender and Grade on Vocab_Score. Be sure to look for any possible interactions before interpreting the main effects.
2. Using the data set Exercise2 in the STATS library, test the effect of Gender and Dose on Flexibility. Again, check for possible interactions. What can you conclude about Gender?
3. Using the data set Interact in the STATS library, run a model with Parts (the number of parts produced per hour) as the dependent variable and Training and Seniority as factors. If there is a significant interaction (of course there is), select the appropriate options to include multiple comparisons for all effects to explain why there is an interaction. (It seems that the younger employees benefit more from training than senior employees.)
4. Using the data set Salary in the STATS library, run a two-way ANOVA with Weekly_Salary as the dependent variable with Age_Group and Education as factors. As always, check the interaction term first. Omit the type I SS and produce all the plots on separate graphs.
5. Repeat exercise 4, except add Gender as a third factor. When you create your model, specify only two-way interactions. (Instead of Full Factorial model, choose N-Way Factorial and enter a 2 in the number of interactions.) Suppress all plots.
6. Using the data set Cars in the SASHELP library, run a two-way ANOVA with Length as the dependent variable, Cylinders and Type as factors. Use the following filter: Cylinders IN (4,6) and Type ne ‘Hybrid’.

11.

Correlation

Introduction

There are several ways to quantify the relationship between two continuous variables, the most common being a Pearson correlation coefficient. This chapter describes this as well as a nonparametric alternative called a Spearman rank-order correlation.

Using the Statistics Correlation Task

As a quick summary, a **Person correlation** coefficient measures the strength of the relationship between two variables. These variables are usually continuous, but there are types of correlations where one or both of the values are binary (0 or 1) or ordinal variables. The formula for calculating Pearson correlation coefficients (from here on, just called correlations) is such that correlation values lie between -1 and +1, inclusive. Positive correlations indicate a positive relationship between two variables. For example, height and weight would be positively correlated for a group of young children. The taller children would tend to be heavier and vice versa. Correlations near zero tell you that given a value for one of the variables, you have no better guess for the value of the other variable. Finally, correlations near -1 indicate a strong inverse relationship between two values; as one increases, the other decreases. For example, the dose of insulin would be negatively correlated with blood sugar levels—the higher the dose, the lower the blood sugar level.

It is important to remember three things about correlations. First, it doesn't matter which variable you place on the X or Y axis. Second, if two variables are correlated, it does not prove causality. Third, the correlation coefficient is strongly influenced by outliers. The following two figures show the effect of a single outlier on a set of data points that, without the outlier, has a correlation close to zero.

First, here is a scatter plot of x-y data with a correlation close to zero.

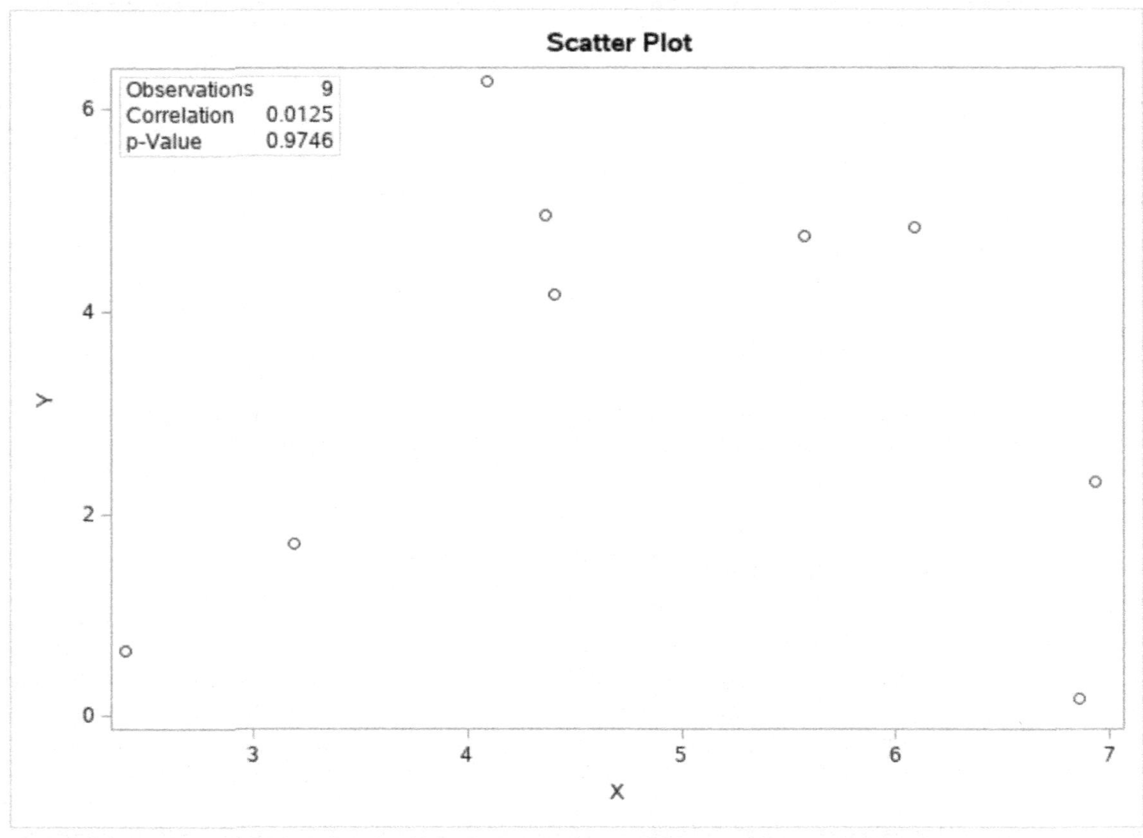

Figure 11.1: X and Y Points without an Outlier

The correlation is .0125 (almost zero). Here is the same set of Xs and Ys with a single outlier (x=14.5, y=19.5) added.

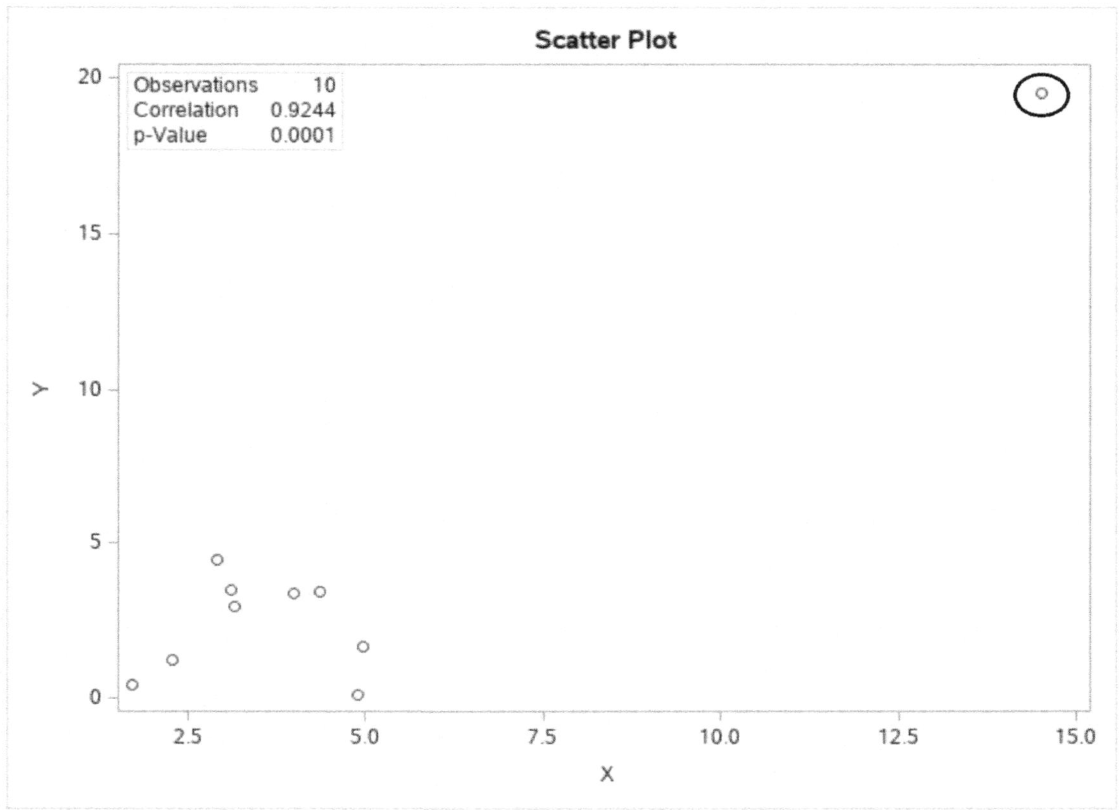

Figure 11.2: Sample Plot with a Single Outlier

The correlation is now .9244. The lesson here is that outliers can have a very large effect on Pearson correlation coefficients. In looking at extreme outliers, you might want to do two things: one is to make sure it isn't a data entry error, and the second is to make sure that this case is really appropriately considered to be part of the population under consideration. For example, if you were looking at the relationship between income and school achievement, and one person in your sample was a multi-billionaire, you might want to remove that data point from the sample. Of course, in doing so, you would want to say that you had done that in any article or report you wrote concerning those data.

> This brings up one of the most important rules about reporting correlation coefficients in a study: always inspect a scatter plot to identify data points that might have undue influence.

It's time to investigate correlations among the variables in the Exercise data set. Start by clicking Tasks and Utilities. Then choose Statistics, and finally, Correlation Analysis:

Tasks and Utilities ⇒ Statistics ⇔ Correlation Analysis

See Figure 11.3 below.

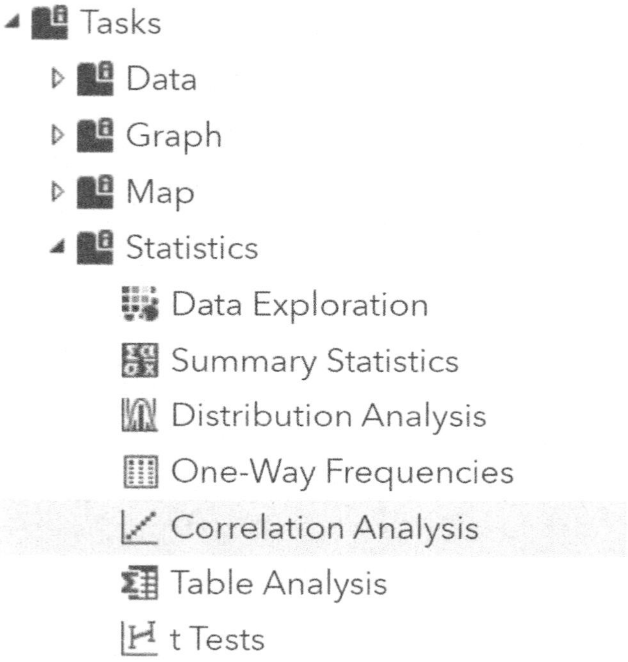

- Tasks
  - Data
  - Graph
  - Map
  - Statistics
    - Data Exploration
    - Summary Statistics
    - Distribution Analysis
    - One-Way Frequencies
    - Correlation Analysis
    - Table Analysis
    - t Tests

Figure 11.3: Selecting the Correlation Task

On the DATA tab, select the Exercise data set in the STATS library. There are two selections under the Roles tab: One is labeled Analysis variables and the other is labeled Correlate with. You have a choice. One option is to enter all the numeric variables of interest in the Analysis variables box (as was done here) to compute correlations between every pair of variables. As an alternative, select one or more variables in the Analysis box and other variables in the Correlate with box. If you do this, the task will compute correlations for every combination of the Analysis variables and the Correlate with variables. For example, if you have variables A and B in the Analysis variables box and variables X, Y, and Z in the Correlate with box, the task will compute the correlations for the pairs: AX, AY, AZ, BX, BY, and BZ.

In this example, you are entering all three variables of interest in the Analysis variables box to produce a correlation matrix for these three variables. (See Figure 11.4.)

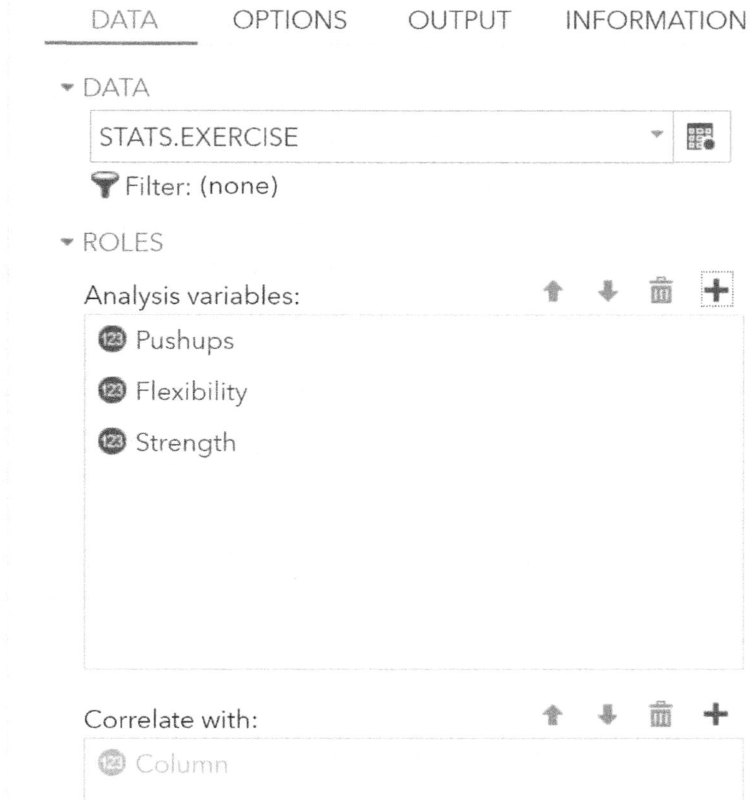

Figure 11.4: The DATA Tab for Correlation Analysis

The next step is to select OPTIONS as described in the section below.

Generating Correlation and Scatter Plot Matrices

Because scatter plots are so important to correlation analysis, select Matrix of scatter plots on the Type of plot menu and check the box to include histograms. The Histograms option adds a histogram for each variable on the main diagonal of the matrix. In the box labeled Display statistics, use the pull-down menu to choose Selected statistics and check the box to Display p-values. These selections are shown in Figure 11.5.

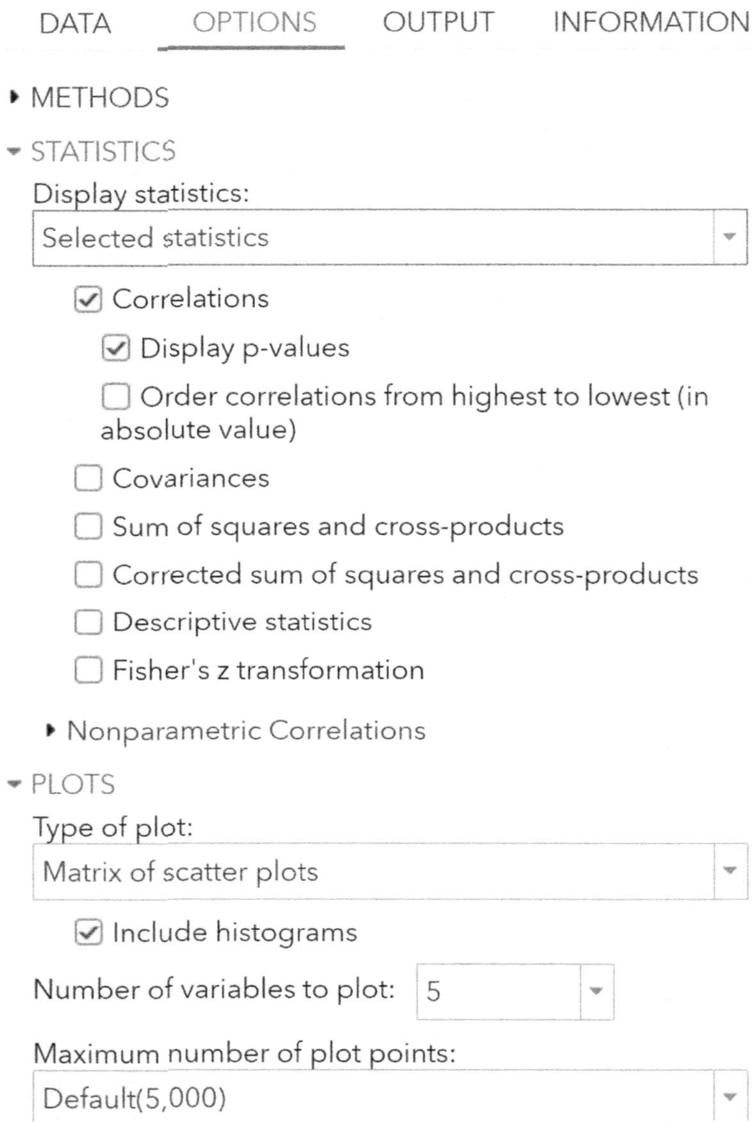

DATA OPTIONS OUTPUT INFORMATION

▸ METHODS

▾ STATISTICS

Display statistics:

Selected statistics

- [x] Correlations
  - [x] Display p-values
  - [ ] Order correlations from highest to lowest (in absolute value)
- [ ] Covariances
- [ ] Sum of squares and cross-products
- [ ] Corrected sum of squares and cross-products
- [ ] Descriptive statistics
- [ ] Fisher's z transformation

▸ Nonparametric Correlations

▾ PLOTS

Type of plot:

Matrix of scatter plots

- [x] Include histograms

Number of variables to plot: 5

Maximum number of plot points:

Default(5,000)

Figure 11.5: The DATA Tab for Correlation Analysis

It's time to run the task. Figure 11.6 shows the first part of the output.

3 Variables:	Pushups Flexibility Strength

Pearson Correlation Coefficients, N = 240 **Prob > |r| under H0: Rho=0**			
	Pushups	**Flexibility**	**Strength**
Pushups	1.00000	0.07060 0.2760	0.22151 0.0005
Flexibility	0.07060 0.2760	1.00000	-0.07764 0.2308
Strength	0.22151 0.0005	-0.07764 0.2308	1.00000

Figure 11.6: Correlation Matrix

The intersection of each row and column in this table shows you the correlation coefficient (top number) and the *p*-value (the bottom number). For example, the correlation between Pushups and Strength is .22151 with a *p*-value of .0005. The more strength a person has, the more push-ups they can perform. Because of the symmetry of the matrix, you only need to look at the upper or lower triangle of the matrix.

Before you spend time investigating the correlation coefficient, you should take a moment to inspect the *p*-value. What does it mean to have a "significant" correlation? To understand the *p*-value, imagine two variables that are completely unrelated—the correlation between them is 0. (See Figure 11.7.)

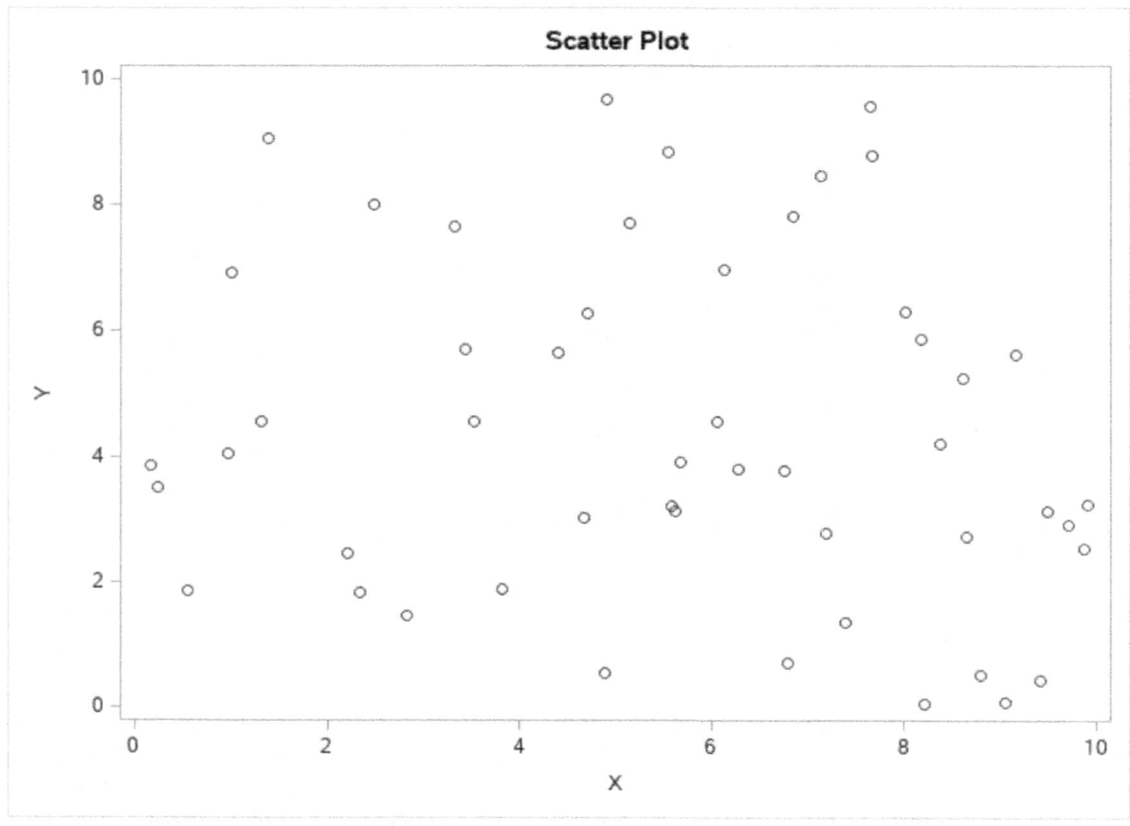

Figure 11.7: Population with a Zero Correlation

Now, imagine taking a random sample of five subjects from this population. You might wind up with a selection that looks like the circled points in Figure 11.8. The correlation, using these five "random" points is .8. The problem is that with a small sample, you may, by chance, end up with a correlation that is either close to 1 or close to -1. The *p*-value that you see in the correlation matrix is the probability that you would obtain a correlation with an absolute value as large as or larger than the one you obtained by chance alone, given that the true population correlation between your two variables is actually 0.

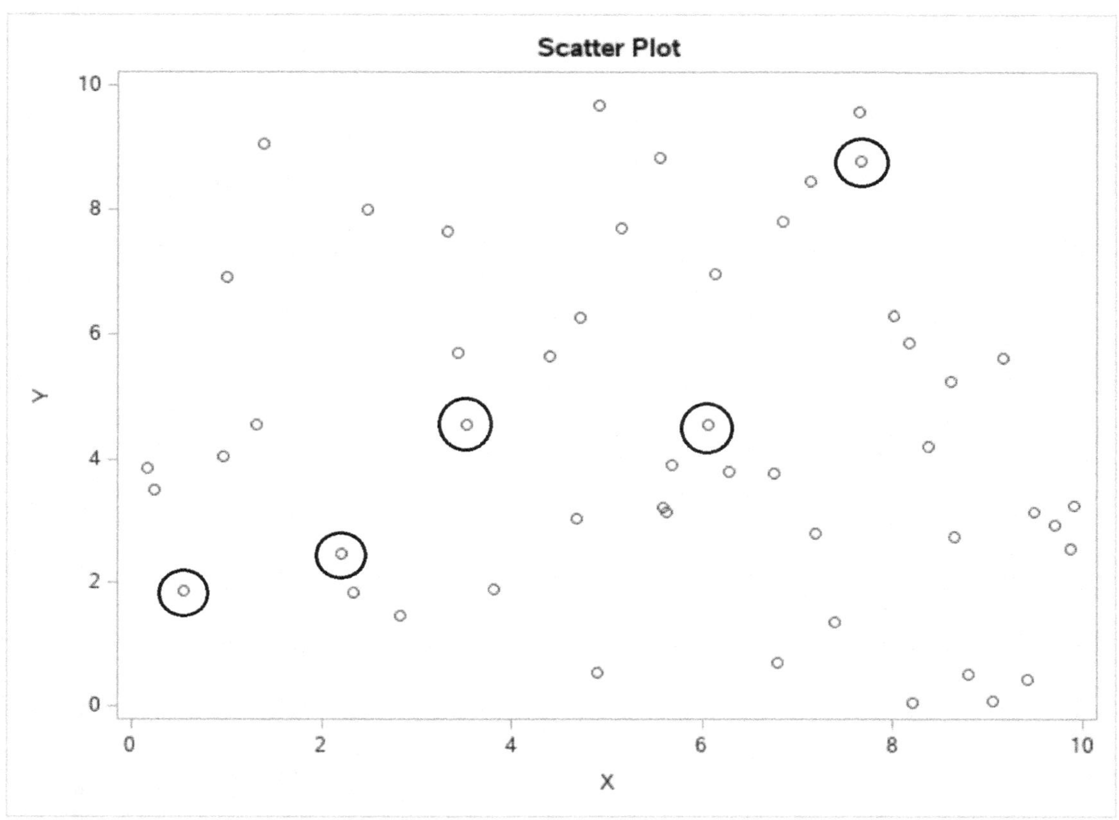

Figure 11.8: Random Sample: Correlation = .8

Now, back to the output. Following the correlation matrix is a matrix of scatter plots. Because you checked the box to include histograms on the diagonal, they are included as well. (See Figure 11.9.)

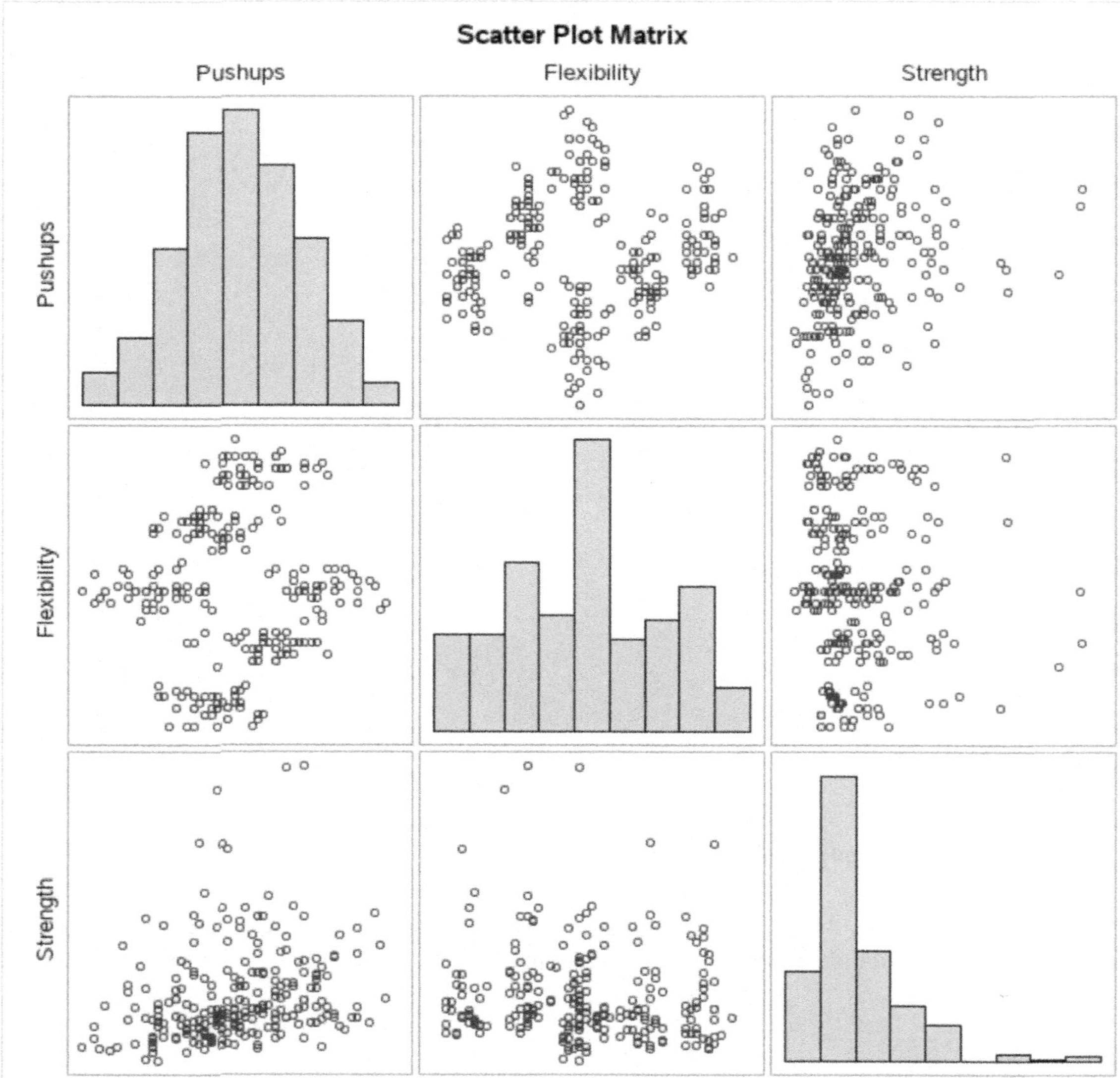

Figure 11.9: Scatter Plot Matrix

Here you see a scatter plot for all combinations of the variables (corresponding to the correlation matrix above). This scatter plot matrix is an excellent way to see relationships among the variables in your data set.

If you choose individual scatter plots instead of a scatter plot matrix, each of the small plots displayed in the matrix is displayed as an individual plot.

Before we move on to interpreting correlation coefficients, let's look at a situation where variables are more highly correlated than in the Exercise data set.

Correlations among Variables in the Fish Data Set

Now that you are an expert in generating correlation matrices, let's correlate several variables in the Fish data set in the SASHELP library. Because there are several different species of fish, you can use a filter to select a single species (Bream) and generate a correlation matrix and a matrix of scatter plots. The DATA tab, with a filter and a list of three variables (Length1, Height, and Width) is displayed in Figure 11.10.

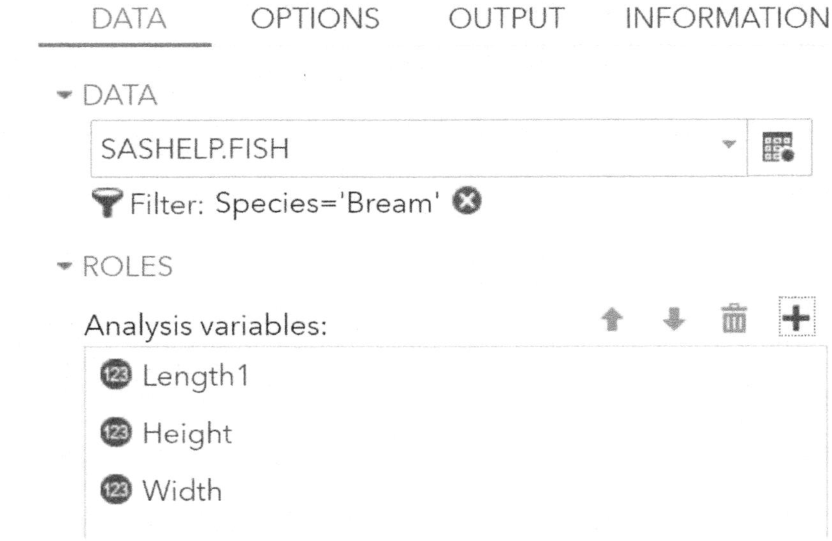

Figure 11.10: Generating Correlations for Bream in the Fish Data Set

Running this task produces the correlation matrix (Figure 11.11) and the scatter plots (Figure 11.12).

3 Variables:	Length1 Height Width

Pearson Correlation Coefficients, N = 35 Prob > |r| under H0: Rho=0			
	Length1	**Height**	**Width**
Length1	1.00000	0.93938 <.0001	0.89930 <.0001
Height	0.93938 <.0001	1.00000	0.92669 <.0001
Width	0.89930 <.0001	0.92669 <.0001	1.00000

Figure 11.11: Correlation Matrix for Three Variables in the Fish Data Set

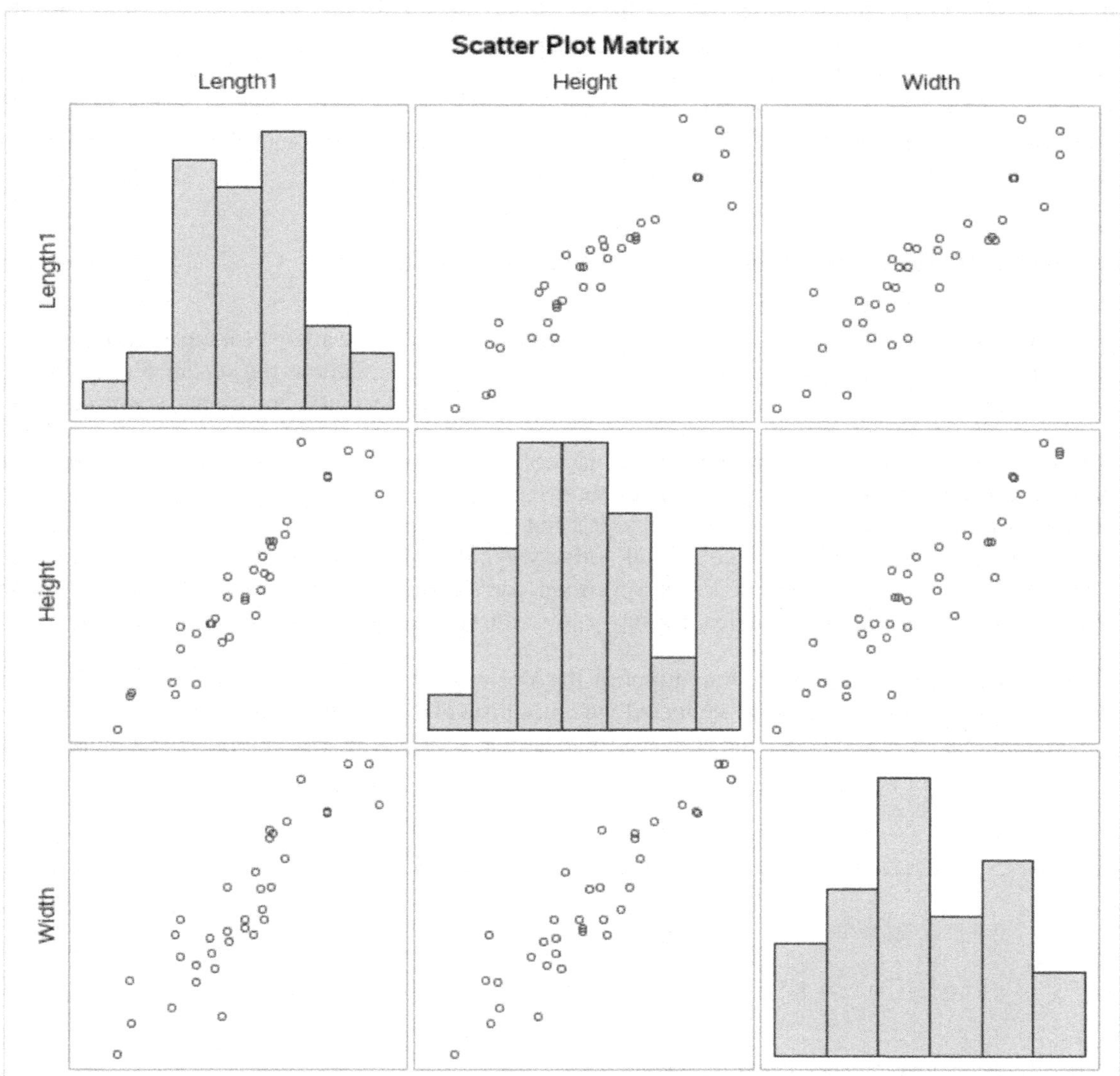

Figure 11.12: Matrix of Scatter Plots

Let's use this matrix to see how to interpret correlation coefficients.

Interpreting Correlation Coefficients

This question often comes up: "What is a large correlation?" The short answer is, "It depends." That's not very satisfying. A useful approach to interpreting a correlation coefficient is to square it. The value of R-square is the proportion of variance (the standard deviation squared) of one of the variables that can be explained by the other. For example, look at the correlation between the variables Height and Width. It is .92669, and this value squared is .85875. Both of these variables (Height and Width) differ among the fish. You can also compute the variance that is jointly shared for the two variables Height and Width. Because the value of R-square is .85875, you can say that about 86% of the variance in width can be explained by the fact that these fish have different heights.

Generating Spearman Nonparametric Correlations

As with many statistical tests, there is a nonparametric alternative to a Pearson correlation. You typically want to run nonparametric correlations when you see outliers in the scatter plots. This was the case in the Exercise data set, so we will use that same data set to run nonparametric correlations.

One of the most popular nonparametric alternatives to a Pearson correlation is called a **Spearman rank-order correlation**. The Spearman method substitutes ranks for the two variables and then computes a correlation based on the ranks. It's not a bad idea to routinely compute both Pearson and Spearman correlations and take special notice when they produce substantially different results. Large differences between Pearson and Spearman correlations usually indicate problems with the distributions of one or both variables. Be sure you create scatter plots to see what is going on.

To add Spearman correlations to the output of the Correlation analysis tab, click the OPTIONS tab, then Statistics. In the menu, choose Selected Statistics instead of Default Statistics. Once there, expand the Nonparametric tab and click the box next to Spearman's rank-order correlation. (See Figure 11.13.)

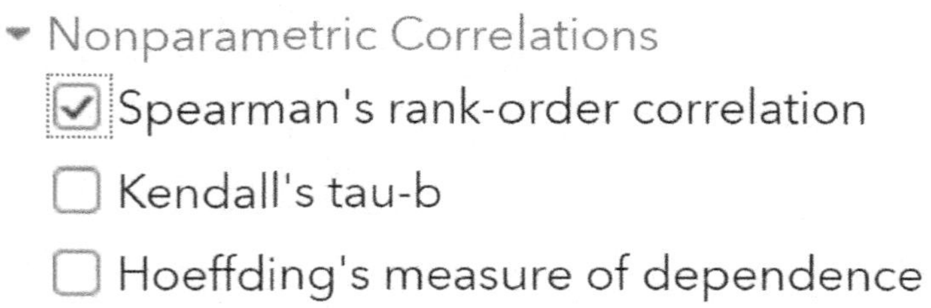

Figure 11.13: Requesting Spearman Correlations

You will now see two correlation matrices—one containing Pearson correlations and one containing Spearman correlations as shown in Figure 11.14.

3 Variables:	Pushups Flexibility Strength

Pearson Correlation Coefficients, N = 240
Prob > |r| under H0: Rho=0

	Pushups	Flexibility	Strength
Pushups	1.00000	0.07060 0.2760	0.22151 0.0005
Flexibility	0.07060 0.2760	1.00000	-0.07764 0.2308
Strength	0.22151 0.0005	-0.07764 0.2308	1.00000

Spearman Correlation Coefficients, N = 240
Prob > |r| under H0: Rho=0

	Pushups	Flexibility	Strength
Pushups	1.00000	0.05841 0.3677	0.30726 <.0001
Flexibility	0.05841 0.3677	1.00000	-0.11347 0.0794
Strength	0.30726 <.0001	-0.11347 0.0794	1.00000

Figure 11.14: Pearson and Spearman Correlations

In this example, the differences between the Pearson and Spearman values are not that large, but it still pays to inspect the scatter plot matrices.

Conclusion

The Correlation analysis statistics task enables you to correlate one set of variables with another or to produce a correlation matrix showing correlations between every pair of variables. You should routinely request either individual plots or a scatter plot matrix as part of the procedure. Finally, consider computing Spearman rank-order correlations, especially when you see outliers in the scatter plots.

Chapter 11 Exercises

1. Using the data set High_School in the STATS library, create a correlation matrix including the variables Vocab_Score, Spelling_Score and English_Grade. Request histograms on the diagonal.
2. Run exercise 1 again, including Spearman nonparametric correlations.
3. Using the Heart data set in the SASHELP library, generate a correlation matrix for the variables Height, Weight, Systolic, and Diastolic. Run it first with no plots. Run it a second time with a request for individual plots. Where are they? Check the Log and repeat the request increasing the number of points from the default of 5,000 to No Limit.
4. Using the High_School data set in the STATS library, generate correlations between the variable English_Grade versus Vocab_Score and Spelling_Score.
5. Using the High_School data set in the STATS library, generate correlations between the variables Vocab_Score, Spelling_Score, and English_Grade with the variable Honor. Because the values of Honor are dichotomous (0 or 1), these correlations are known as point-biserial correlations. Suppress all plots.
6. First, run the short program below to create a WORK data set called XY. Then use the Correlation task to generate both Pearson and Spearman correlations. Include a plot. Notice how one outlier can affect these two values:

```
data XY;
   call streaminit(13579);
   do i = 1 to 9;
      x = round(rand('uniform')*100);
      y = round(.75*x + rand('uniform')*10);
      output;
   end;
   x = 95;
   y = 5;

   output;
run;
```

7. Using the data set Physics_Test in the STATS library, compute the point-biserial correlations between the variable Grade (grade on the test) and each of the 10 answer variables (Ans1-Ans10). (Hint: Enter Grade as your Analysis variable and the 10 variables Ans1-Ans10 as your Correlate with variables.) Suppress all plots. The point-biserial correlation coefficient is one way that psychometricians determine how well an item is performing on a test. Very low values or even negative values indicate a problem with the item.

12.

Simple and Multiple Regression

Introduction

Correlation analysis identifies relationships between variables—regression analysis enables you to predict one variable based on values of another variable (simple regression) or a combination of variables (multiple regression).

The concept behind simple or multiple regression is to select one or more variables (sometimes called predictor variables) that can be combined to predict an outcome (dependent variable). The models described in this book are all linear models.

There are several reasons why regression techniques are useful. One reason is that you can use the resulting regression equation to predict a value. For example, given the gender, height, and weight of a person, you could predict the maximum lung volume for that person. Another reason to use multiple regression is to help explain relationships among a set of variables. That is, you can look at the influence of one or more variables to better understand the nature of a dependent variable. If more than one predictor variable is used, the analysis looks at all predictor variables simultaneously.

Regression, especially multiple regression, needs to be done carefully. There are a number of diagnostic plots produced by the Regression task that are useful in deciding if the assumptions for regression are satisfied by your data. In addition, you can test for influential data points (data values that can change the results of the analysis if they are included or not), and test for multi-collinearity (predictor variables correlated with each other). Including highly correlated predictor variables in a multiple regression can lead to strange and paradoxical results.

Getting an Intuitive Feel for Regression

Before we get into the details of running regression models, let's look a simple example. If you measured the height and weight of children, you would expect a relationship between these two variables. As you saw in the last chapter, if you computed a correlation coefficient between height and weight, you would expect a positive value. It might look like the plot shown in Figure 12.1.

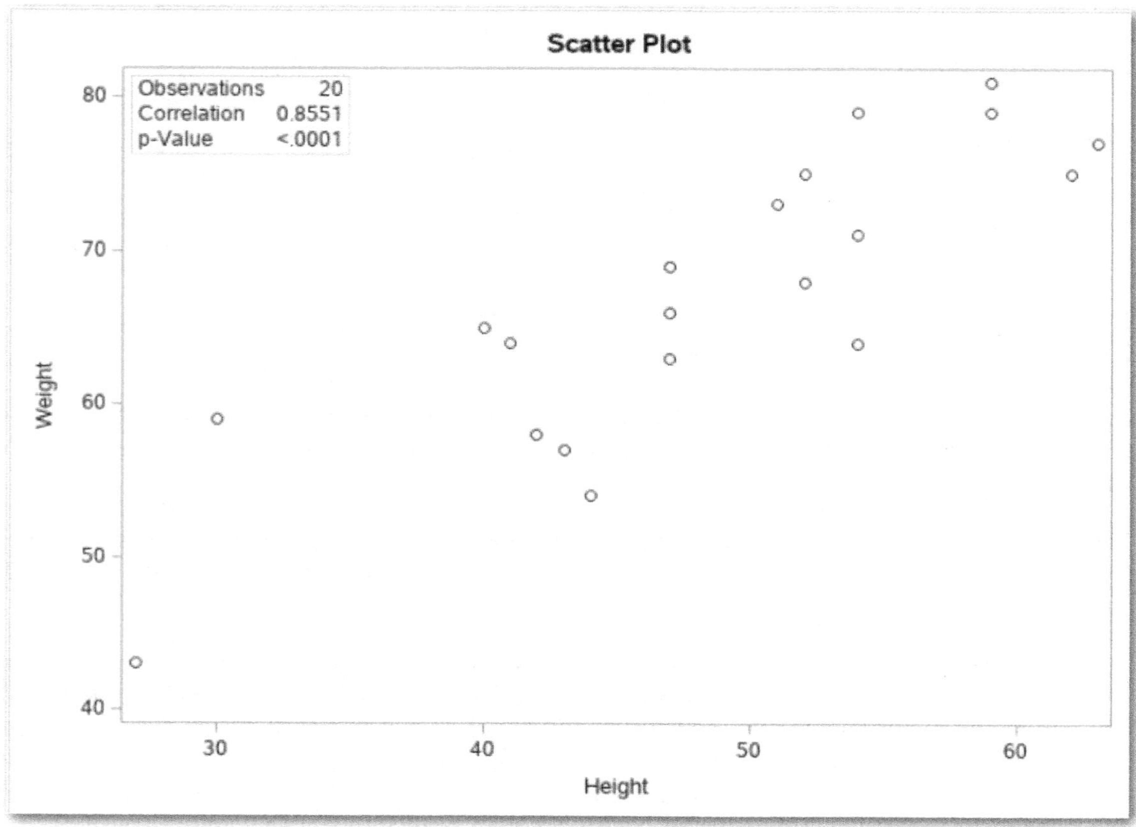

Figure 12.1: Scatter Plot of Height versus Weight

There seems to be a relationship between height and weight—the taller a person is, the heavier. Notice that the correlation is .8551 and it is significant. What is the "best" straight line that you can draw through these data points? One way to determine how to draw the "best" line is move a ruler around until half of the data points fall above the line and half of the data points fall below the line. Another method considers the distance between the line (called a **regression line**) and each data point (this distance is called a **residual**). Two residuals are identified in Figure 12.2. The regression methods in this chapter choose a line where the sum of the squared residuals is minimized. Let's see what such a line looks like.

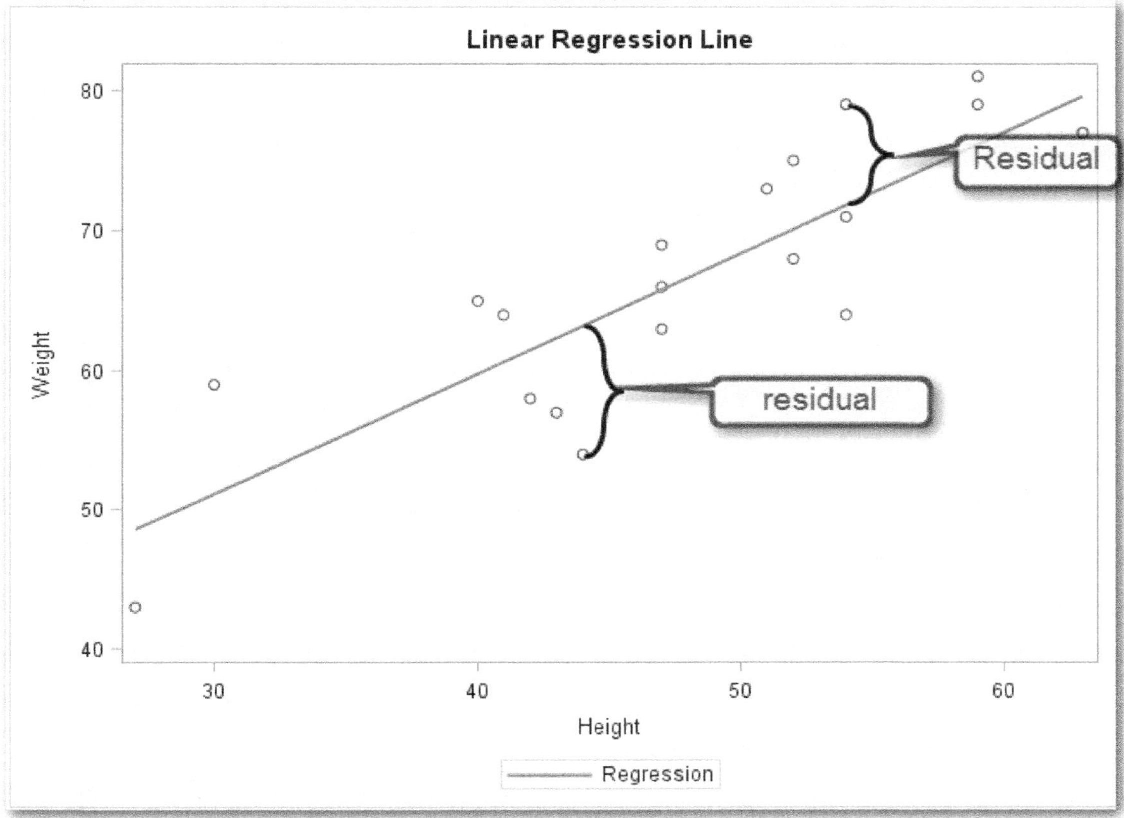

Figure 12.2: Using Linear Regression to Find the "Best" Line

Once you have an equation for this line, you can predict the weight of a person, given the person's height. Later in this chapter, you will see how you can use several variables to predict a value.

Describing Simple Linear Regression

Examples in this chapter use a data set called Exercise2. Some of the variables in this data set are also found in the Exercise data set used in the last chapter. However, to demonstrate some aspects of multiple regression, some new variables were added, and some relationships were modified. We will start by running a simple linear regression to predict the number of push-ups a person can do (variable Pushups) and a variable that represents a person's strength (variable Strength). To start, select:

Tasks and Utilities ⇒ Linear Models ⇒ Linear regression

It looks like this:

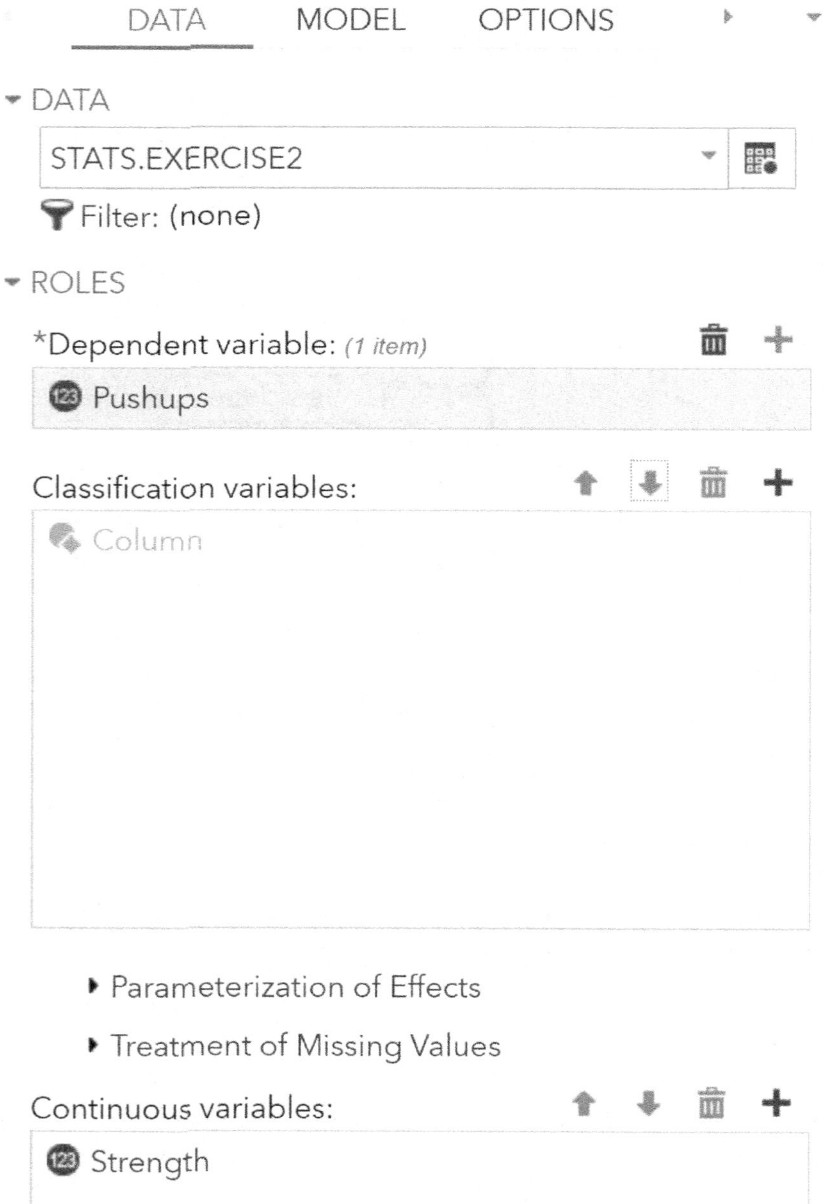

Figure 12.3: Select Data Source and Specify Variables

Select the Exercise2 data set from the STATS library and choose Pushups as the Dependent variable and Strength as the Continuous variable.

The next step is to specify your model. Click the Model tab, then Edit to do this:

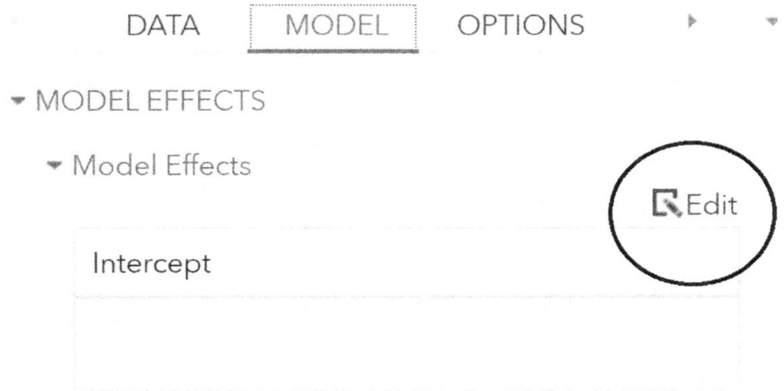

Figure 12.4: Specify the Model

When you click the Edit icon to, the Model Effect Builder appears. It looks like the figure below.

Figure 12.5: Specifying Your Model

Under Variables on the left side of the screen, highlight the variable or variables that you want in your model. In this example, click the variable Strength to select it. Next, click the box labeled "Add."

Your model is shown under the heading "Model effects" on the right side of the screen.

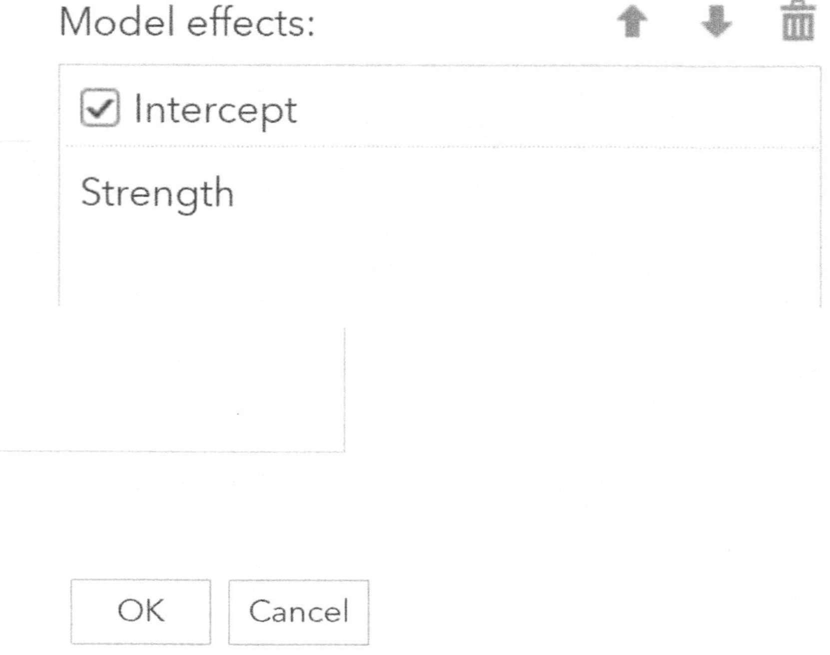

Figure 12.6: Listing the Variables in Your Model

Click OK to complete your model. (You might have to scroll down to the bottom of the screen to see the OK box.)

At this point, you can run the model or click the OPTIONS tab to select additional statistics and plots. In this example, the only option that was changed was to generate all the plots as separate plots (so that the screen shots could be displayed better).

Model: MODEL1
Dependent Variable: Pushups

Number of Observations Read	120
Number of Observations Used	120

Analysis of Variance					
Source	**DF**	**Sum of Squares**	**Mean Square**	**F Value**	**Pr > F**
Model	1	5993.96562	5993.96562	140.06	<.0001
Error	118	5049.90105	42.79577		
Corrected Total	119	11044			

Root MSE	6.54185	**R-Square**	0.5427
Dependent Mean	33.13333	**Adj R-Sq**	0.5389
Coeff Var	19.74401		

Parameter Estimates					
Variable	**DF**	**Parameter Estimate**	**Standard Error**	**t Value**	**Pr > |t|**
Intercept	1	11.22108	1.94545	5.77	<.0001
Strength	1	0.59356	0.05015	11.83	<.0001

Figure 12.7: First Section of Linear Regression Output

Because the number of observations read and used are equal (with a value of 120), you know that there were no missing values for the variables Pushups and Strength. The mean square due to the model is much larger than the mean square due to error, yielding a very large F value and a low *p*-value. The mean of the dependent variable (Pushups) is 31.13333, and the R-square is .5427. The Linear regression task also computes an adjusted R-square. This value is useful when you have several independent variables in the model. The addition of independent variables in a model causes the value of R-square to increase even if the independent variables are only randomly correlated with the dependent variable. The value labeled Adj R-Sq adjusts for the number of independent variables in your model and is a way to compare models with different numbers of independent variables.

Near the bottom of the figure, you see the intercept and slope of the regression line. If you wanted to compute the number of push-ups a person could do, given a value for strength, you could use the following linear equation:

Pushups = 11.22108 + .59356*Strength

For example, if a person's strength was 10, you would estimate that they could do about 17 push-ups.

Before you start interpreting the relatively high R-square and the low *p*-value for this model, you need to look at some diagnostic plots to determine if the residuals are normally distributed and if the variance of the residuals seems homogeneous across different values of Strength. These are two assumptions behind linear regression. It's time to inspect some of the diagnostic plots.

Figure 12.8 is one of the diagnostic plots produced by the regression task. It shows the difference between the actual value of Pushups and the value predicted by the regression.

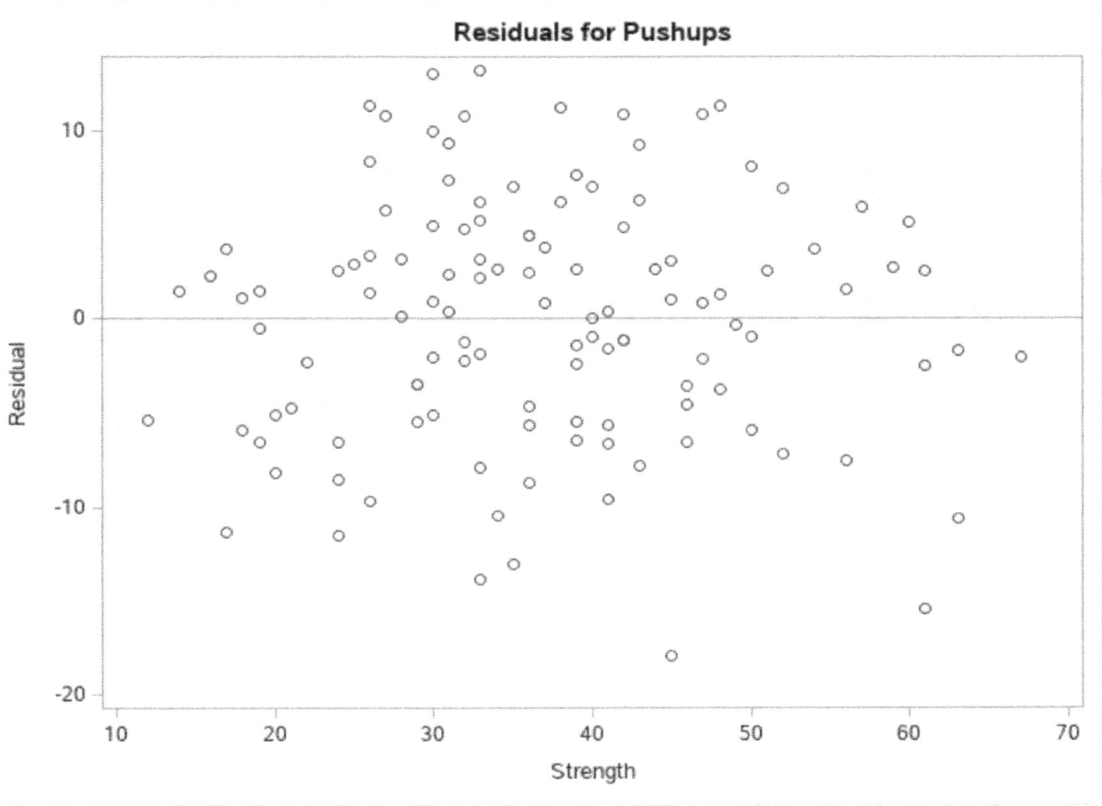

Figure 12.8: Residual Plot

The points on this plot seem pretty random—this is a good thing. If there appears to be a pattern (such as a curve) in this plot, you might need to consider a non-linear term in the regression. Notice that as you go from left to right in the plot, the amount of scatter above and below the line does not change too much. This is also a good thing, suggesting that the variance of Pushups is homogeneous for different values of Strength.

Understanding How the F Value Is Computed

To help explain the two sources of variation in a simple linear model, a single data point was selected in Figure 12.9 (located in the upper right side of the plot). The horizontal line on the plot is the mean number of Pushups for the total sample (about 33 push-ups). You can compute the difference of each data point from this mean value. If you take this difference, square it, and add up all the squared values, you will be computing a value known as the **sum of squares** (SS). The selected data point contributes to the total sum of squares for two reasons. First, given a value of Strength equal to about 50 (the value of Strength for the selected point), you would expect the number of push-ups to also be about 50 (the value that lies on the regression line). The difference between the mean number of push-ups and the point on the regression line represents the amount that the regression contributes to the sum of squares for this particular point. The difference between the regression line and each data point (called a residual) contributes to the error sum of squares. If all the data points were close to the regression line, most of the contributions to the sum of squares would be due to the regression, and a smaller portion would be due to error, indicating a good fit. The F value is computed from these two components of the total sum of squares.

If you divide the sum of squares by the degrees of freedom (shown in Figure 12.7 as DF), you get the values labeled Mean Squares, which is another name for variance. The ratio of the mean square due to the model divided by the mean square due to error (computed from the residuals), gives you the F value shown in the table. In the old days (pre-computer or in a class where the professor made you compute everything by hand), you could look up a probability of getting an F value this large of larger by chance alone. By using SAS Studio, you can see the *p*-value as less than .0001 (highly significant).

The shaded portion of this graph represents the 95% confidence limit for the prediction of Pushups for any given value of Strength. The other, much wider confidence limits, indicated by the dotted lines on the plot, represent a confidence interval for individual data points. Given a value of Strength, you are 95% confident that an individual data point will fall within these limits.

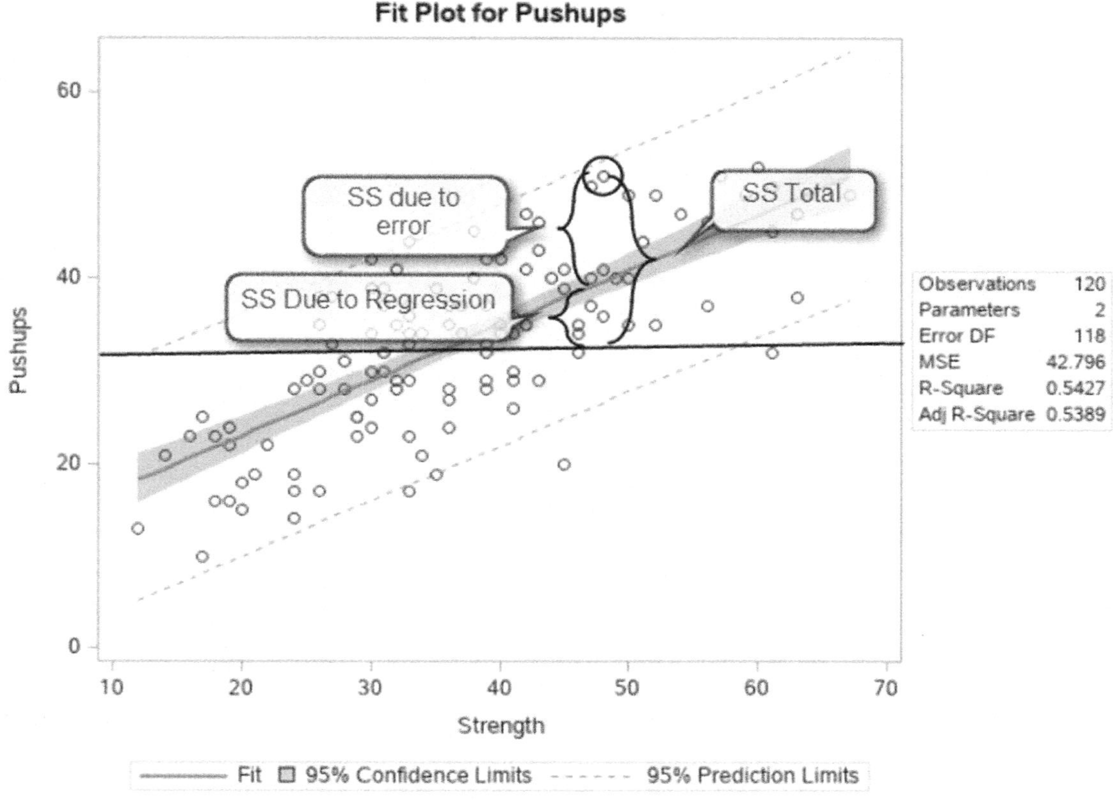

Figure 12.9: Fit Plot for Pushups

Investigating the Distribution of the Residuals

Even though the residual plot looks "pretty good" to the naked eye, it is still a good idea to use one of the diagnostic plots to check how much the distribution deviates from a normal distribution. Figure 12.10 shows a histogram of the residuals with a superimposed normal distribution. This should give you confidence that the requirements for fitting a regression model were satisfied.

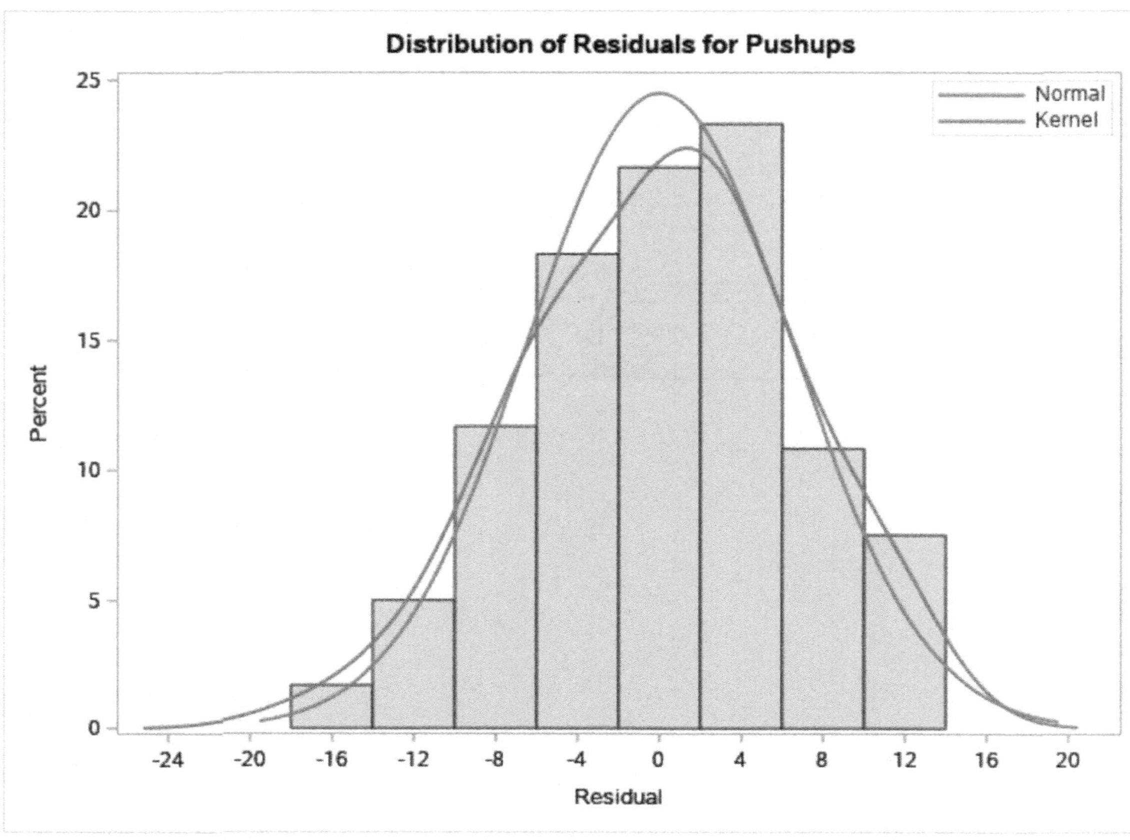

Figure 12.10: Checking the Distribution of the Residuals

Measures of Influence

Certain data points (such as outliers) can have a profound effect on both correlation and regression analysis. There are several measures of influence available in the regression task. They all involve running the regression with and without each data point to see if the removal of a particular data point changes either the predicted value or the betas (coefficients) in the regression equation. One popular measure of influence is called **Cook's D**. Cook's D measures changes in the predicted value when a data point is removed. Large values for Cook's D indicate data points that are influential.

To have the Linear Regression task output Cook's D and other measures of influence, first click the OPTIONS tab. Next, in the box below Display Statistics, use the menu to change Default Statistics to Default and Selected Statistics. Finally, check the box labeled Analysis of Influence (one of the selections listed under Diagnostics).

Figure 12.11 shows the value of Cook's D for each observation in the data set. If you place your cursor on one of these large values, the observation number for that data point is displayed (not shown in this screenshot). You might want to check that this data point is valid and not the result of a data error.

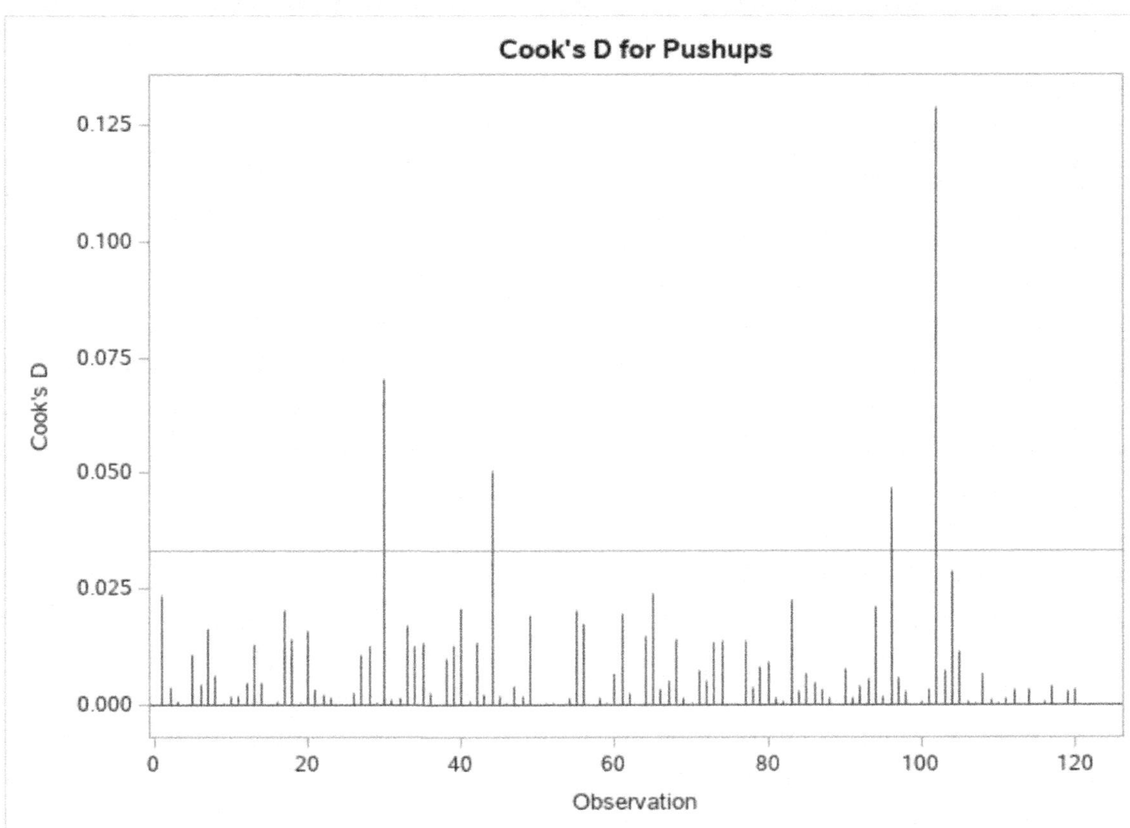

Figure 12.11: Using Cook's D as a Measure of Influence

If each observation in your data set has a unique identifier (such as Subject or ID), you can have that value added to Observation number in the Cook's D output. To accomplish this, notice the three tabs at the top of the results window on the right side of your screen. They are labeled CODE, LOG, and RESULTS. If you click the CODE tab, you will see all the code that was generated by the Linear Regression task. Next, click Edit. This enables you to modify the generated code. If you add an ID statement such as:

```
id Subject;
```

before or after the MODEL statement and rerun the model, you will see the value of the variable Subject in the box that shows the value of Cook's D and the Observation number.

Demonstrating Multiple Regression

Multiple regression, as the name implies, uses more than one predictor variable to estimate a single dependent variable. You can use the same Exercise2 data set used in the simple linear regression section to see how to run a multiple regression using SAS Studio. A useful first step is to use the Correlation task to produce a correlation matrix for all the variables of interest.

As you saw in the previous chapter, you need to open the Correlation Task, enter all the variables of interest, request a matrix of scatter plots (check the box to include histograms in the OPTIONS window), and run the task. Here are the results:

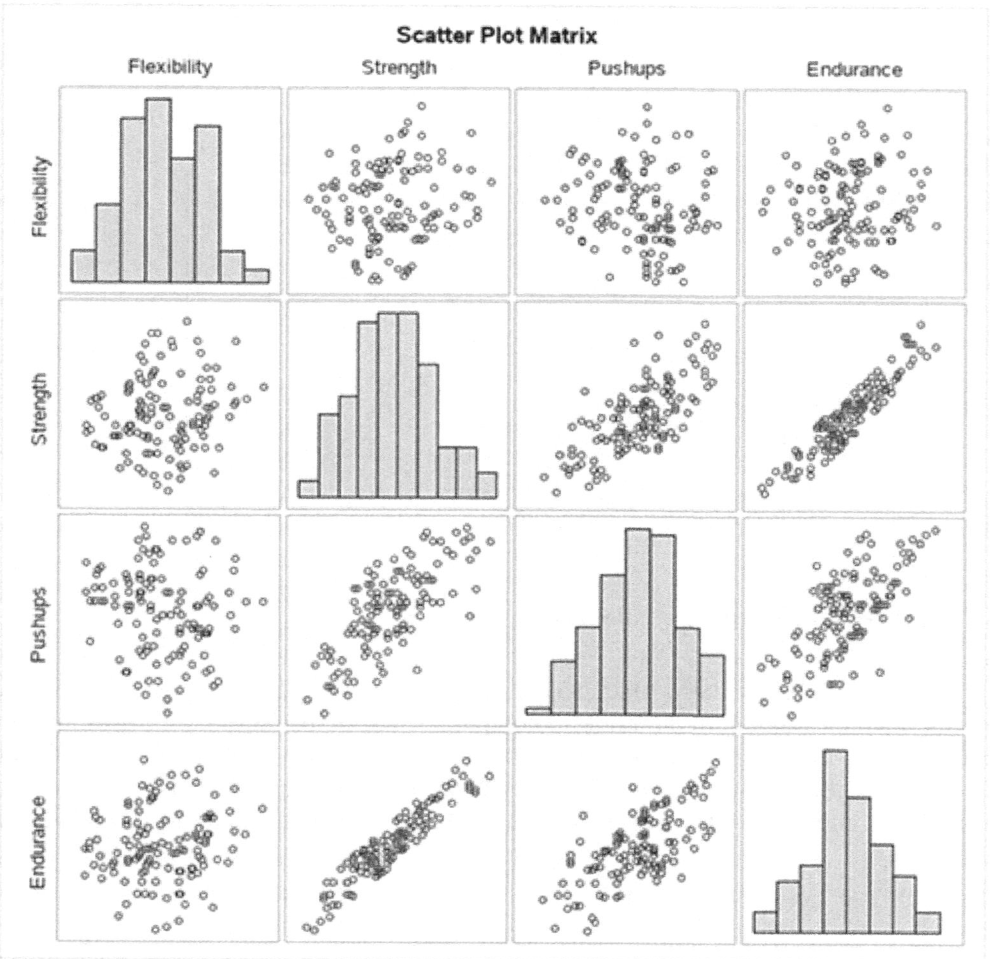

Figure 12.12: Matrix of Scatter Plots from Exercise2 Data Set

Some of the variables such as Strength and Endurance are highly correlated—others such as Flexibility and Strength are not. The condition where the predictor variables are highly correlated is called multi-collinearity, and it causes serious problems when these variables are used together in multiple regression (as you will see in the section on collinearity).

Running a Simple Linear Regression Model with Endurance and Pushups

Because Pushups and Endurance are highly correlated (the correlation coefficient is .69228), you would expect that you could predict the number of push-ups a person could do by knowing the value of the variable Endurance. Let's check if this is true.

Running a model with Pushups as the dependent variable and Endurance as the independent (predictor) variable results in the following output.

Model: MODEL1
Dependent Variable: Pushups

Number of Observations Read	120
Number of Observations Used	120

Analysis of Variance					
Source	DF	Sum of Squares	Mean Square	F Value	Pr > F
Model	1	5292.94023	5292.94023	108.60	<.0001
Error	118	5750.92644	48.73666		
Corrected Total	119	11044			

Root MSE	6.98116	R-Square	0.4793
Dependent Mean	33.13333	Adj R-Sq	0.4749
Coeff Var	21.06991		

Parameter Estimates					
Variable	DF	Parameter Estimate	Standard Error	t Value	Pr > |t|
Intercept	1	9.77952	2.32983	4.20	<.0001
Endurance	1	0.53178	0.05103	10.42	<.0001

Model: MODEL1
Dependent Variable: Pushups

Figure 12.13: Using Endurance to Predict the number of Push-ups

The R-square is .4793, and the coefficient for Endurance is highly significant. Because Pushups and Strength are also correlated (r = .73671), you could repeat the Linear regression task using Strength instead of Endurance to predict the number of push-ups a person could do. If you ran that model, you would see an R-Square of .5427 and a *p*-value that is highly significant ($p < .0001$). The coefficient of Strength (the slope) is also positive and equal to .59356.

What happens if you include both Strength and Endurance in a single model? That is the topic of the next section.

Demonstrating the Effect of Multi-Collinearity

To demonstrate what happens when you include two or more highly correlated independent variables in a multiple regression model, let's include both Strength and Endurance ($r = .93681$) in a model to predict the number of push-ups a person can do. Output from a model with both Strength and Endurance as independent variables is shown in Figure 12.14.

Model: MODEL1
Dependent Variable: Pushups

Number of Observations Read	120
Number of Observations Used	120

Analysis of Variance					
Source	DF	Sum of Squares	Mean Square	F Value	Pr > F
Model	2	5994.37687	2997.18843	69.45	<.0001
Error	117	5049.48980	43.15803		
Corrected Total	119	11044			

Root MSE	6.56948	R-Square	0.5428
Dependent Mean	33.13333	Adj R-Sq	0.5350
Coeff Var	19.82740		

Parameter Estimates					
Variable	DF	Parameter Estimate	Standard Error	t Value	Pr > |t|
Intercept	1	11.11869	2.21746	5.01	<.0001
Endurance	1	0.01340	0.13726	0.10	0.9224
Strength	1	0.58039	0.14397	4.03	<.0001

Figure 12.14: Output with Both Strength and Endurance as Predictor Variables

In this model, Strength is highly significant, but Endurance is not. The coefficient for Endurance is close to zero and the *p*-value is .9224. What happened?

The variable Strength had strong predictive value in predicting the number of push-ups a person can do. Because it explains almost all of the variance of the dependent variable, Endurance does not have much more to contribute to the model. Remember that you are looking for the unique contribution of each variable after all the other variables have been accounted for. The correlation matrix shown earlier told you that Strength and Endurance were highly correlated, and it would not be wise to include both of them in a single model.

When more than two variables are highly correlated, inspecting the correlation matrix might not always point out a multi-collinearity problem. The SAS Studio Linear Regression task includes several ways to detect multi-collinearity. One of the most popular (and easiest to understand) methods is called the variance inflation factor (abbreviated VIF).

Here's how VIF works: If you take each of the predictor variables one at a time and perform a multiple regression with the selected predictor variable as the dependent variable and the remaining predictor variables as independent variables, you will get an R-square for each predictor variable. High values of R-square indicate that a particular predictor variable is correlated with a linear combination of the other predictor variables. Think of it as the degree to which the variable under consideration is more or less already accounted for by the other variables in the regression. Rather than use the R-square values, the VIF is computed as follows:

$$VIF_i = \frac{1}{(1 - R_i^2)}$$

For example, if an R-square is equal to .9, the VIF will be 1/(1-.9) = 10. Therefore, large values of VIF indicate multi-collinearity problems.

To include the VIF in your regression output, click the OPTIONS tab, expand the menu next to Display statistics, and select Collinearity. Check the box for Variance inflation factors (Figure 12.15).

▾ Collinearity

☐ Collinearity analysis

☐ Tolerance values for estimates

☑ Variance inflation factors

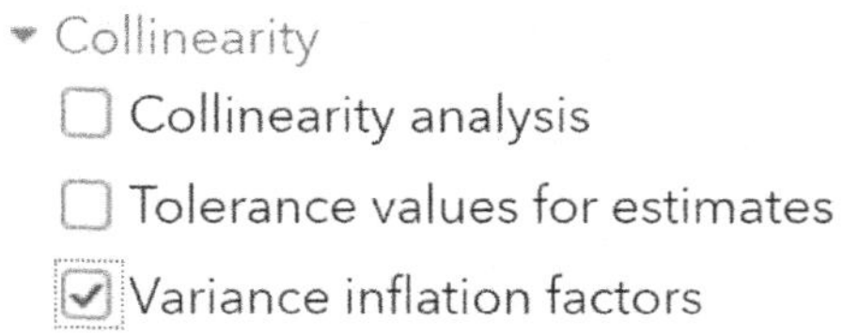

Figure 12.15: Requesting the VIF

Let's run the task with predictor variables (Endurance, Strength, and Flexibility). The VIF is now included in the output. (See Figure 12.16.)

Parameter Estimates						
Variable	**DF**	**Parameter Estimate**	**Standard Error**	**t Value**	**Pr > |t|**	**Variance Inflation**
Intercept	1	18.56557	2.82929	6.56	<.0001	0
Endurance	1	-0.01533	0.12977	-0.12	0.9062	8.19665
Strength	1	0.63531	0.13662	4.65	<.0001	8.25758
Flexibility	1	-0.71620	0.18307	-3.91	0.0002	1.02380

Figure 12.16: Output Including the VIF

As expected, the VIFs for Endurance and Strength are high. Your job now is to either combine these two variables into a single index or to eliminate one of them from the model. The decision of which variable to remove is not always straightforward. Many researchers remove the variable with the largest *p*-value. However, in actual practice, you might choose a different variable because it is easier or cheaper to measure.

The next section of output (with Endurance removed) shows that collinearity is no longer a problem. The adjusted R-square is slightly higher (.5891) than it was in the model with only Strength included (adjusted R-square was .5389). You might decide that this very slight increase in R-square is not worth adding Flexibility to the model. That is a judgment call.

Analysis of Variance					
Source	DF	Sum of Squares	Mean Square	F Value	Pr > F
Model	2	6582.40607	3291.20303	86.31	<.0001
Error	117	4461.46060	38.13214		
Corrected Total	119	11044			

Root MSE	6.17512	R-Square	0.5960
Dependent Mean	33.13333	Adj R-Sq	0.5891
Coeff Var	18.63719		

Parameter Estimates						
Variable	DF	Parameter Estimate	Standard Error	t Value	Pr > |t|	Variance Inflation
Intercept	1	18.43605	2.59724	7.10	<.0001	0
Strength	1	0.62020	0.04783	12.97	<.0001	1.02052
Flexibility	1	-0.71497	0.18200	-3.93	0.0001	1.02052

Figure 12.17: Output from a Model with Endurance Removed

Demonstrating Selection Methods

The Linear Regression task has a SELECTION tab that gives you a choice of several different methods of automatically selecting a model. (See Figure 12.18.)

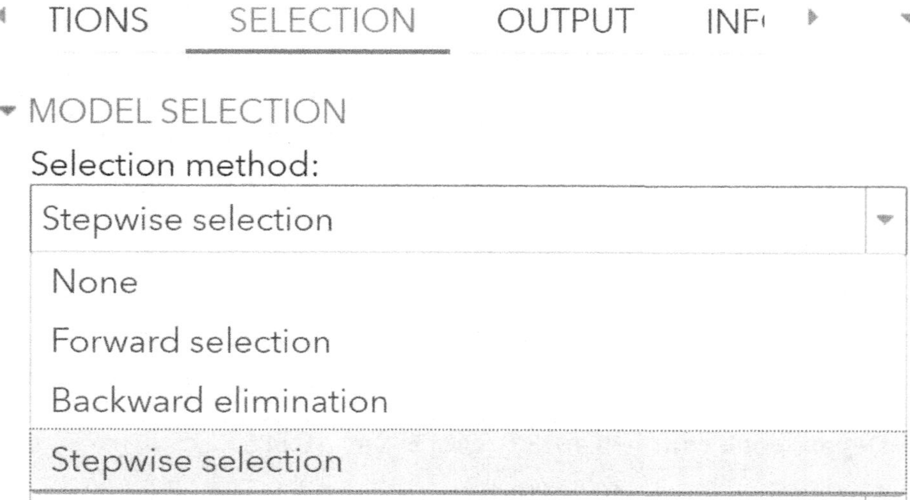

Figure 12.18: Options on the SELECTION Tab

If you leave the default value of None, the regression task will enter all of the predictor variables that you have selected, either as Classification variables or Continuous variables. The other three methods in the list represent automatic selection methods.

Each of these automatic selection methods enables you to choose a criterion for entering or removing variables.

- **Forward selection** first chooses the "best" variable depending on which of the selection criteria you have chosen. For example, if you chose R-square as your selection criteria, the variable with the highest R-square value (lowest *p*-value) would enter first. Forward selection would continue to enter the remaining variables based on values of R-square until the *p*-value for a variable exceeds a default value.
- **Backward selection** is just the opposite—it starts with all the variables in the model and removes them one-by-one until a criterion is met.
- One of the most popular selection methods is the last one in the list—**stepwise selection**. This method is similar to forward selection, except that variables already selected can be removed if adding other variables causes a particular variable to no longer satisfy the selection criterion.

Once you choose a selection method, you need to specify a criterion for adding or removing predictor variables in your model. The default method is called the **Schwarz Bayesian Information Criterion (SBC).** This method aims to find a model with good predictive properties while limiting the number of predictor variables in the model (called a parsimonious model). Another popular selection method uses the value of R-square or the adjusted R-square to determine which variables to enter or remove from the model.

To demonstrate an automatic selection process, let's use all the continuous variables in the Exercise2 data set and request stepwise regression. Let's also use the default add and remove criteria SBC.

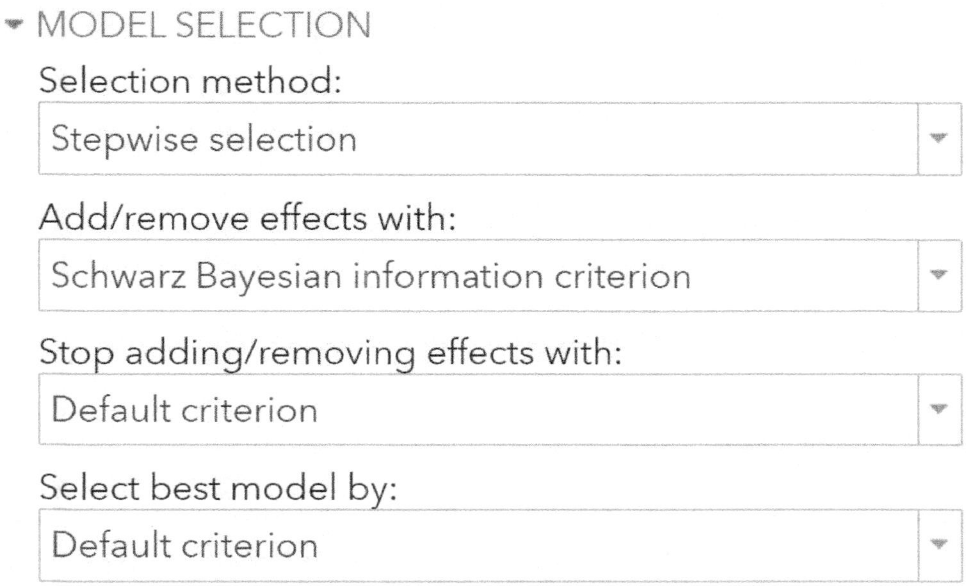

MODEL SELECTION

Selection method:

Stepwise selection

Add/remove effects with:

Schwarz Bayesian information criterion

Stop adding/removing effects with:

Default criterion

Select best model by:

Default criterion

Figure 12.19: Selecting Stepwise Regression and Schwarz Bayesian Information Criterion

Here is the first portion of the output:

Stepwise Selection Summary				
Step	Effect Entered	Effect Removed	Number Effects In	SBC
0	Intercept		1	547.4441
1	Strength		2	458.3308
2	Flexibility		3	448.2512*
* Optimal Value of Criterion				

Selection stopped because all effects are in the final model.

Figure 12.20: Output Using the Stepwise Method

Not surprisingly, the stepwise method chose Strength and Flexibility as predictor variables. The next figure shows four different criteria for deciding when to stop adding variables.

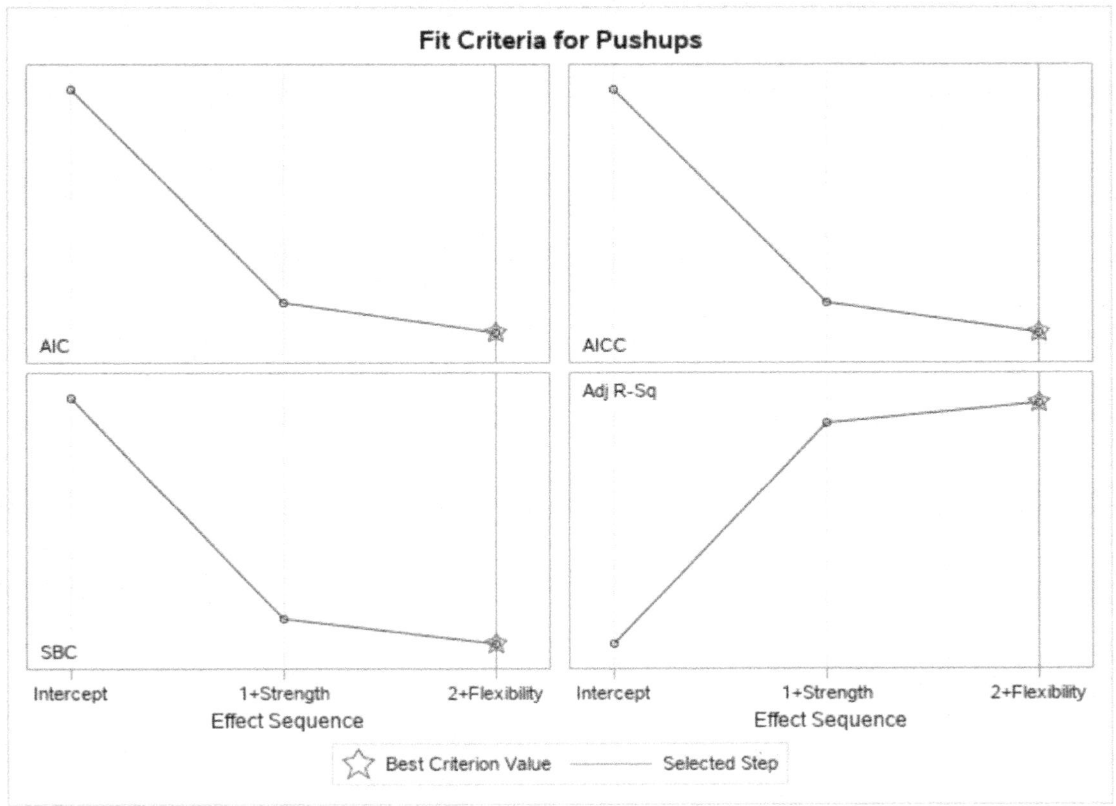

Figure 12.21: Four Methods for Determining When to Stop Adding Variables

AIC (Akaike's information criteria – pronounced A-KA-ee-kay) and a modified version, AICC, are displayed in the top two graphs. The stars on the plots indicate that the two-variable model is the "best" model as defined by each criterion. The SBC, discussed earlier, is displayed on the bottom left, and the adjusted R-square is displayed on the bottom right. Although all four criteria selected the same model, that is not always the case.

> The order of entry using any of the automatic selection methods does not necessarily tell you the relative importance of each variable in predicting the dependent variable. For example, you could have a variable that is highly correlated with the dependent variable but does not enter the model because it is also correlated with other predictor variables that entered previously.

Using a Categorical Variable as a Predictor in Model

The Regression task also enables you to select categorical variables, such as Gender, in your model. Let's run a regression model with Gender and Strength as the predictor variables. The DATA tab looks like this.

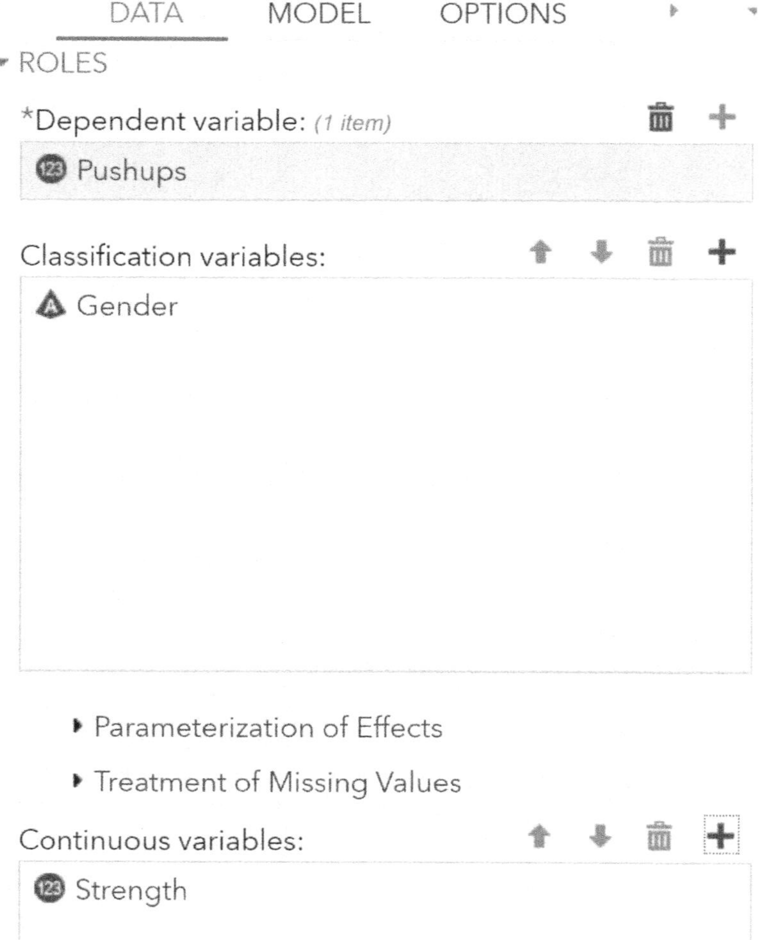

Figure 12.22: Selecting a Categorical Variable in a Regression

Behind the scenes, the task is creating a dummy variable with values of 0 and 1 to represent the genders M and F. Here is the first section of the output:

Least Squares Model (No Selection)

Analysis of Variance					
Source	DF	Sum of Squares	Mean Square	F Value	Pr > F
Model	2	7433.78401	3716.89200	120.46	<.0001
Error	117	3610.08266	30.85541		
Corrected Total	119	11044			

Root MSE	5.55476
Dependent Mean	33.13333
R-Square	0.6731
Adj R-Sq	0.6675
AIC	536.47930
AICC	536.82713
SBC	422.84178

Figure 12.23: Output Using Gender as a Predictor Variable

You see a highly significant *p*-value (<.0001) and an R-Square of .6731. The next portion of output shows the coefficients for Gender and Strength in the regression equation.

Parameter Estimates					
Parameter	DF	Estimate	Standard Error	t Value	Pr > |t|
Intercept	1	22.233220	2.308152	9.63	<.0001
Strength	1	0.406769	0.050610	8.04	<.0001
Gender F	1	-8.232902	1.205216	-6.83	<.0001
Gender M	0	0	.	.	.

Figure 12.24: Parameter Estimates for Gender and Strength

Both Strength and Gender are significant, and the R-Square is higher than any of the models you have run so far. The task chose M as the reference value and the estimate of -8.232902 for the coefficient means that, on average, women did that many fewer push-ups compared to men.

Conclusion

As demonstrated in this chapter, failing to pay attention to the diagnostic information, especially multi-collinearity, can result in models that do not make sense. Besides checking the residuals to see if they are approximately normally distributed and that the variance of the residuals is somewhat homogeneous for different values of your independent variables, there is one additional caution. There is a "rule of thumb" that says you should have approximately 10 times the number of observations as predictor variables in your model. If you have too few observations, you might see very high values of R-square that are simply a mathematical fluke and not actual relationships in your model.

It is probably better to use a rule of thumb that says something more like: 50 observations plus 10 times the number of independent variables.

Chapter 12 Exercises

1. Using the data set High_School in the STATS library, run a simple linear regression with Spelling_Score as the dependent variable and Vocab_Score as the independent (continuous) variable. What is the R-square? What is the predicted value of spelling score for a value of 100 on the vocabulary score?
2. Continuing with the High_School data set, run a simple linear regression using

Vocab_Score to predict English_Grade. Run the model again adding Spelling_Score as a continuous predictor variable. Notice the betas (coefficients) and run a test for collinearity using the variance inflation factor (VIF).

3. Repeat the second part of exercise 2 and add the categorical variable Honor (1=in the honor society, 0=is not in the honor society) and Gender to the model.
4. Repeat exercise 3, but use stepwise model selection. Did the adjusted R-square increase? Sometimes fewer variables make for a better model. Use the default SBC stopping criteria. Rerun the model using R-square instead of SBC. Any differences in the model? Change the criteria to adjusted R-square. The fact that adding predictor variables to models increases R-square is one reason that it might be preferable to use adjusted R-square.
5. Using the data Cars in the SASHELP library, run a multiple regression with MSRP as the dependent variable and the three variables Horsepower, Weight, and Length as predictor variables. Use the stepwise selection method and the default Schwarz Bayesian Criterion (SBC) as the method for adding or removing effects.
6. Repeat exercise 5, except do not use a selection method. Include a request to compute the VIF (variance inflation factor) as a measure of collinearity. Notice the *p*-value for Weight. Why is it not significant? Create a correlation matrix and a matrix of plots for the four variables MSRP, Horsepower, Weight, and Length. This should help you understand why this is always a good step before you dive into multiple regression analysis.

13.

Binary Logistic Regression

Introduction

In the last chapter, you learned how to create multiple regression models. Conceptually, logistic regression has some similarities to multiple regression, although the computational method (maximum likelihood) is quite different (and CPU-intensive). Multiple regression uses a set of predictor variables to predict and model a continuous outcome variable. **Binary logistic regression** uses a set of predictor variables to predict a dichotomous outcome. Theoretically, a multiple regression equation can predict values from negative infinity to positive infinity—binary logistic regression is attempting to compute a probability that an event occurs or does not occur. Because probabilities are bounded between 0 and 1, multiple regression should not be used. Instead, a transformation (called a logit) is performed so that the results of a binary logistic model are bounded by 0 and 1. The transformation, for the mathematically interested reader, is to take the natural log of the odds (the probability that the event occurs divided by the probability that the event does not occur). Luckily, the results of a binary logistic model provide you with odds ratios (OR) for classification variables that have a straightforward interpretation.

Describing the Risk Data Set

Binary logistic regression uses a set of predictor variables to predict a dichotomous outcome. It is possible to perform multi-nominal (more than two outcomes) using logistic regression models. However, the interpretation of such models is quite complicated and beyond the scope of this book. Therefore, the example that follows investigates well-known risk factors such as age, cholesterol, and gender to predict if a person is going to have a heart attack. Although the risk factors are in line with current thinking about heart disease, they were simulated (and exaggerated) for this data set.

Variables in the data set are shown in the following table.

Variable	Description
Gender	Coded as M for Male and F for Female
Age	Age in years
Age_Group	Three groups: 1:Less 60, 2:60–70, 3:Over 70
Chol	Total Cholesterol
Chol_High	A dichotomous character variable: 0=normal, 1=high
Heart_Attack	A dichotomous numeric outcome variable: 1=yes, 0=no

Running a Binary Logistic Regression Model with a Single Predictor Variable

The first step is to select the Binary Logistic task in the Linear Models menu. (See Figure 13.1.)

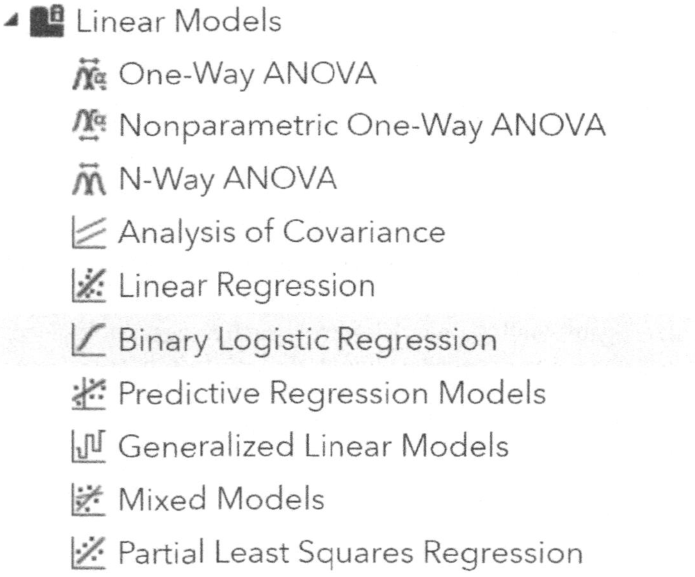

Linear Models
- One-Way ANOVA
- Nonparametric One-Way ANOVA
- N-Way ANOVA
- Analysis of Covariance
- Linear Regression
- Binary Logistic Regression
- Predictive Regression Models
- Generalized Linear Models
- Mixed Models
- Partial Least Squares Regression

Figure 13.1: Selecting Binary Logistic Regression Task

This brings up the DATA tab where you get to choose the data set, the response variable (the variable that you are trying to predict), and two categories of predictor variables: continuous or categorical. We will start out with a single categorical variable (Gender) for this first example.

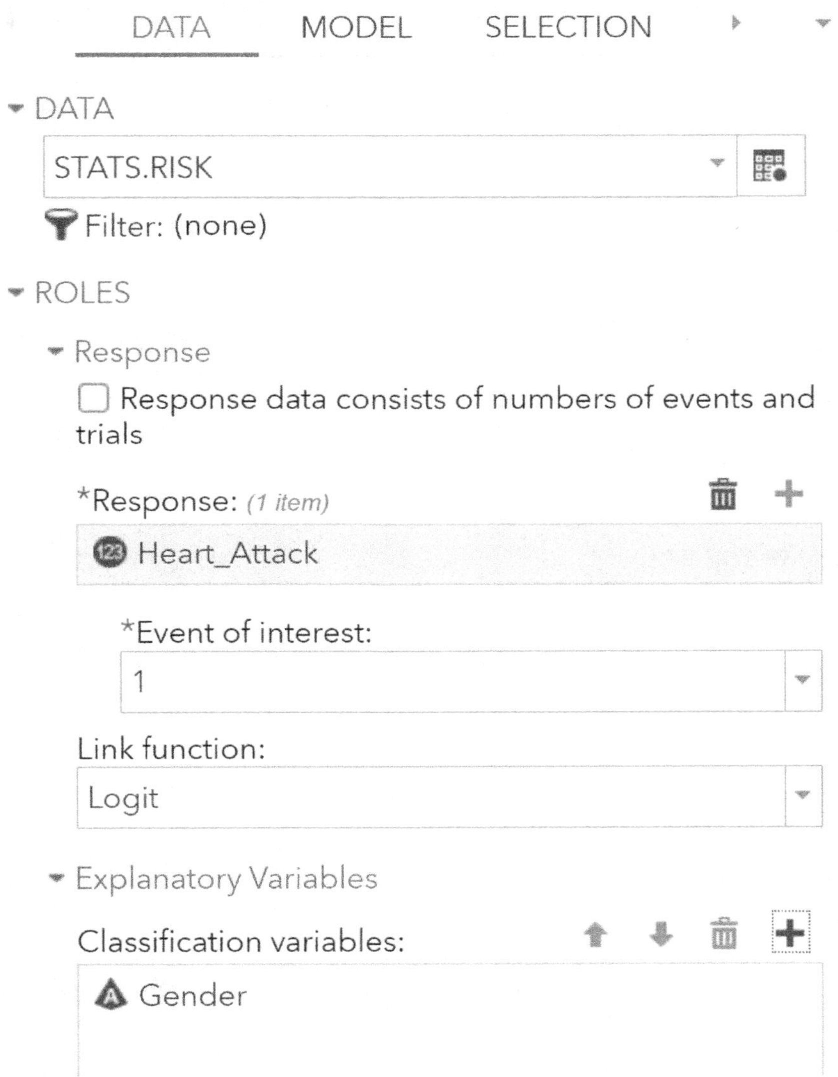

DATA MODEL SELECTION

DATA

STATS.RISK

Filter: (none)

ROLES

Response

Response data consists of numbers of events and trials

*Response: *(1 item)*

Heart_Attack

*Event of interest:

1

Link function:

Logit

Explanatory Variables

Classification variables:

Gender

Figure 13.2: Selecting the Data Set and One Predictor Variable

In Figure 13.2, showing the DATA tab, you are choosing the Risk data set (in the STATS Library), Heart_Attack as the response variable, and Gender as the Classification variable. You also need to select a value of interest for the Response variable. If you click the down-arrow at the right side of the Response box, you can select one of the two values from the list. In this example, the choices are 1 (yes) and 0 (no). Because you want to look at risks for having a heart attack, you select a value of 1.

Toward the bottom of the DATA tab, look for the menu under Parameterization of Effects and select Reference coding. This method is the most common coding method and, at least to this author, the easiest to interpret. It selects one level of a classification variable as a reference level, and each of the odds ratios is computed by comparing a particular category to this reference level (more on this later).

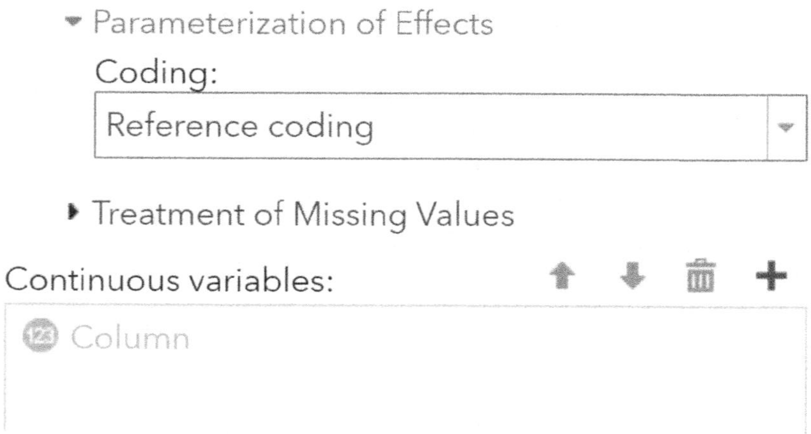

Parameterization of Effects
Coding:
Reference coding
Treatment of Missing Values
Continuous variables:
Column

Figure 13.3: Selecting the Coding Method

It's time to move to the MODEL tab (Figure 13.4).

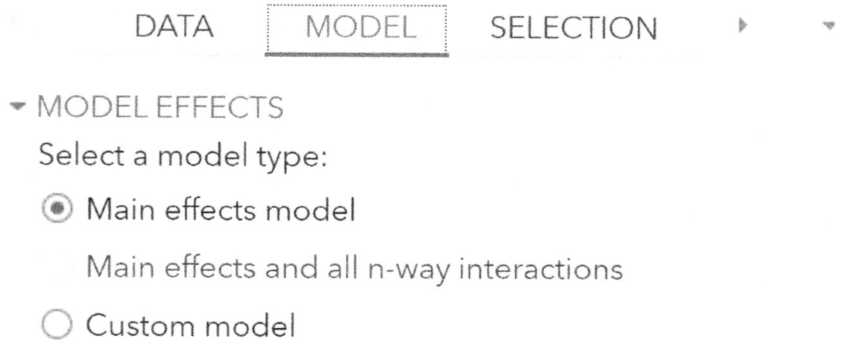

DATA MODEL SELECTION
MODEL EFFECTS
Select a model type:
Main effects model
Main effects and all n-way interactions
Custom model

Figure 13.4: Options on the MODEL Tab

Because this model has only one predictor variable, leave the default selection of Main effects model. Later on, when you have two or more predictor variables, you have a choice of using only main effects or including interactions. It's time to run the task. Figure 13.5 shows the first part of the output.

Model Information	
Data Set	STATS.RISK
Response Variable	Heart_Attack
Number of Response Levels	2
Model	binary logit
Optimization Technique	Fisher's scoring

Number of Observations Read	500
Number of Observations Used	500

Figure 13.5: First Part of the Output from the Logistic Task

This is similar to the output that you saw in the regression models in the last chapter. As always, be sure to look at the number of observations read and used. Because there are no missing values in the Risk data set, these two numbers are equal. Next, you see the Class Level Information.

Class Level Information		
Class	Value	Design Variables
Gender	F	1
	M	0

Model Convergence Status
Convergence criterion (GCONV=1E-8) satisfied.

Figure 13.6: Class Level Information

Values of Gender in the data set are F and M. The binary logistic task has recoded these values to 1 (female) and 0 (male). In the "old days," SAS programmers had to create these design variables themselves—now this task is done for you. Underneath the Class Level Information, you see a message that the convergence criterion was satisfied. That is a good thing. If you see any other message here, go and seek help from your friendly neighborhood statistician.

The next section of output is labeled Fit Statistics. These are a bit complicated and are more useful when comparing models. Smaller values of AIC (Akaike's information criteria) indicate better models. The criterion labeled SC (**Schwarz Criteria**) is based on the value of AIC but adjusts for the number of variables in the model. As with AIC, smaller values of SC indicate better models. The Schwarz Criterion is probably better to use than AIC if you want a parsimonious model (fewer predictor variables—fewer is usually better).

Model Fit Statistics		
Criterion	**Intercept Only**	**Intercept and Covariates**
AIC	360.879	351.339
SC	365.093	359.768
-2 Log L	358.879	347.339

Testing Global Null Hypothesis: BETA=0			
Test	**Chi-Square**	**DF**	**Pr > ChiSq**
Likelihood Ratio	11.5397	1	0.0007
Score	11.2342	1	0.0008
Wald	10.5994	1	0.0011

Figure 13.7: Fit Statistics

At the bottom of Figure 13.7, you see three tests of the null hypothesis. If these are not significant, it might be back to the drawing board. (Do younger readers even know what a drawing board is?) Here three tests of the null hypothesis all reject it with very low *p*-values.

A Discussion about Odds Ratios

Before we discuss odds ratios, it would be a good idea to define odds. Most of us are familiar with probabilities. For example, we all have an intuitive feel for a statement such as "there is a 75% chance that it will rain tomorrow." Statisticians usually express probabilities as decimals, so the

75% probability of rain would be stated as p = .75. As you read in the introduction to this chapter, probabilities must be between 0 and 1 (inclusive). Unless you bet on horse races, you might not be as familiar with **odds**.

Probabilities are computed as the number of events, called "successes," (even if the event is death or some bad outcome) divided by the total number of trials. For example, if you had a group of 100 people and you predicted that 75 of them would be cured of a certain disease within a year, the probability of being cured would be 75/100 or .75. Odds are computed as the number of successes divided by the number of failures. In the case of the cure rate for this disease, the odds of being cured are the number of predicted cures (75) divided by the number of expected failures (25), giving you the odds of 75/25 or 3 to 1. You interpret this to mean that for every three successes, you expect one failure.

In horse racing, odds are posted as the number of losses compared to the number of wins (we will not get into true odds versus payout odds here). For example, if the odds for a horse are posted as 5 to 1, that means that in 6 races, the horse would lose 5 times and win once. The probability of losing would be 5 divided by 6 or about .83.

Mathematically, probability can be computed as:

```
p = Number of successes / Total number of trials
```

Odds can be computed as:

```
Odds = Number of successes / Number of failures
```

A little math should convince you that, given a value for odds, you can compute a probability as:

```
p=Odds/(Odds+1)
```

In the cure rate above where the odds were 3 to 1, the probability would be 3 / (3 + 1) = .75.

We are now ready to talk about odds ratios. Suppose the odds of people with normal cholesterol having a heart attack in the next year are estimated to be .1, and the odds of people with high cholesterol having a heart attack in the next year are estimated at .12. The ratio of these odds is called an **odds ratio**, and it is defined as the odds of having a bad outcome such as a heart attack in the risk group (high cholesterol) divided by the odds of having the same bad outcome in a group not at risk. You could say that the odds ratio for high cholesterol related to having a heart attack is .12 / .1 or 1.2. In other words, a person with high cholesterol is 1.2 times more likely to have a heart attack compared to a person with normal cholesterol. You could also express this result by saying you have a 20% higher risk of having a heart attack if you have high cholesterol.

> Before we go any further into this discussion and investigate and interpret the output from the logistic regression, it should be pointed out that a ratio of incidence rates is called the relative risk. However, in cases where the incidence is relatively low, the odds ratio is a good estimate of relative risk.

You are now ready to understand the next section of output from the Binary logistic regression task (Figure 13.8). You see that the odds ratio for being female (compared to being male, with an outcome of heart attack) is .372. This means it is less likely for a female to have a heart attack compared to a male. (Note: in the past, many women were underdiagnosed for heart attacks. The results presented here are from simulated data to demonstrate binary logistic regression and do not reflect reality.)

Odds Ratio Estimates			
Effect	Point Estimate	95% Wald Confidence Limits	
Gender F vs M	0.372	0.205	0.675

Figure 13.8: Odds Ratios

Because an odds ratio is just what the name implies (a ratio), you could take the reciprocal (1 / .372 = 2.69) and conclude that men are 2.69 more likely to have heart attacks than women. You will see shortly how to tell what value you would like to use as a reference level rather than have SAS Studio do it for you.

One measure of how well your binary logistic equation is predicting the correct outcome is referred to as concordant and discordant pairs listed in Figure 13.9.

Here is an explanation of concordant and discordant pairs: You start by taking all possible pairs of subjects in which one has an outcome in question (heart attack) and the other does not. In this data set there were 58 subjects who had heart attacks and 442 who did not. Taking all possible pairs of subjects results in 58 x 442 = 25, 636 pairs. For each of these pairs, compute the probability of each person in the pair having a heart attack, based on the model. If the prediction is in the same direction as the actual pair (that is, the person with the higher probability is the one who had the heart attack), then the pair is labeled **concordant**. If not, the pair is labeled **discordant**. If the probabilities are computed to be the same (because this example has only the variable Gender, the probabilities will be the same if both subjects in the pair have the same gender) the pair is labeled **tied**. You would like the percent of concordant pairs to be higher than the percent of discordant pairs (as is the case here).

Association of Predicted Probabilities and Observed Responses			
Percent Concordant	37.3	Somers' D	0.234
Percent Discordant	13.9	Gamma	0.458
Percent Tied	48.9	Tau-a	0.048
Pairs	25636	c	0.617

Figure 13.9: Concordant and Discordant Pairs

Editing SAS Studio-Generated Code

Looking back at the odds ratio for Gender, it would be easier to interpret if we made women (Gender = F) the reference value and compared the odds for men (Gender = M) having a heart attack compared to women. To do this, you need to modify the SAS code generated by the SAS Studio task. That is one of the great things about SAS Studio—you can let it do almost all of the work and then edit the code and modify it to your liking. The first step is to click the CODE tab and click the Edit button.

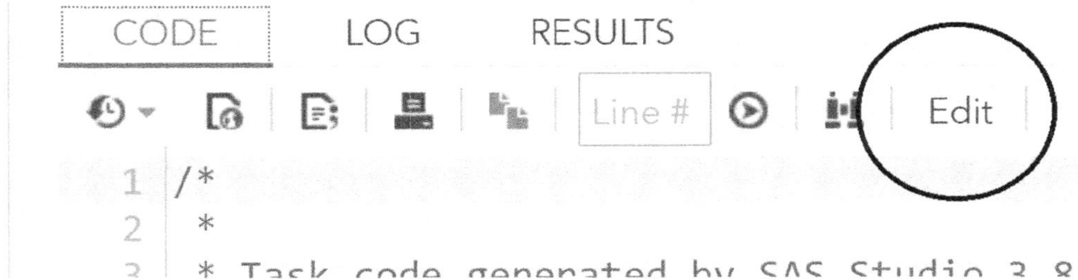

Figure 13.10: Editing the SAS Studio-Generated Code

Next, add the option (ref = 'F') in the CLASS statement as shown in Figure 13.11.

```
proc logistic data=STATS.RISK;
   class Gender (ref='F') / param=ref;
   model Heart_Attack(event='1')=Gender / link=logit technique=fisher;
run;
```

Figure 13.11: Editing the SAS Code to Make Females the Reference Level

If you now rerun the task, the Odds ratio estimates are for men compared to women. (See Figure 13.12.)

Odds Ratio Estimates			
Effect	**Point Estimate**	**95% Wald Confidence Limits**	
Gender M vs F	2.689	1.482	4.877

Figure 13.12: Odds Ratio Estimates with Females as the Reference Level

When you use the REF= option in the CLASS statement, the value that you supply is always in single or double quotation marks, even if the variable is numeric.

Using a Continuous Variable as a Predictor in a Logistic Model

The example shown next uses the variable Chol (cholesterol) as a predictor in a model with heart attack as the outcome variable (and does not include Gender). You start out as before by opening the DATA tab on the Binary Logistic Regression tab. Next, select the Risk data set in the STATS library as your data set. Instead of selecting a Classification variable, select Chol (cholesterol) in the box labeled Continuous variables. (See Figure 13.13 below.)

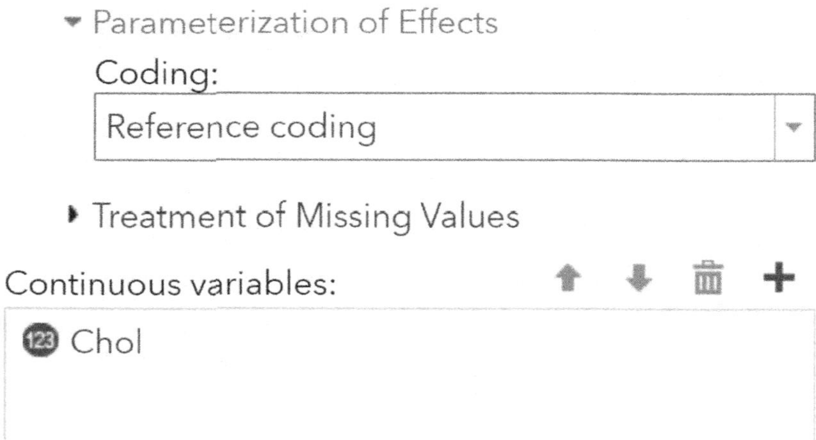

Figure 13.13: Using a Continuous Variable as a Predictor

On the MODEL tab, create your model in a similar manner to the previous example where you were modeling a Classification variable. When you run this task, the first portion of the output helps you decide whether you should reject the null hypothesis of no relationship.

Testing Global Null Hypothesis: BETA=0			
Test	**Chi-Square**	**DF**	**Pr > ChiSq**
Likelihood Ratio	16.8901	1	<.0001
Score	16.1835	1	<.0001
Wald	15.4836	1	<.0001

Figure 13.14: Testing the Null Hypothesis

Because all three measures are highly significant, you conclude that there is a relationship between cholesterol and having a heart attack. The odds ratio estimate is shown in Figure 13.15.

Odds Ratio Estimates			
Effect	**Point Estimate**	**95% Wald Confidence Limits**	
Chol	1.019	1.010	1.029

Figure 13.15: Point Estimate for One Unit of Cholesterol

The value of 1.019 is the risk associated with a single unit of cholesterol. It would be more meaningful to see an odds ratio for a larger value of cholesterol, such as 10 or 20. To add that capability to the procedure, you need to edit the code generated by the binary logistic task, just as you modified the reference level for Gender earlier in this chapter. Open the CODE window and click the Edit button as before, and add a UNITS statement as shown in Figure 13.16.

```
proc logistic data=STATS.RISK;
   model Heart_Attack(event='1')=Chol / link=logit technique=fisher;
   units Chol=10;
run;
```

Figure 13.16: Editing the Code by Adding a UNITS Statement

Now, when you run the task, you get all of the output that you got before, plus the odds ratio associated with a change of 10 units of cholesterol (Figure 13.17).

Odds Ratios		
Effect	**Unit**	**Estimate**
Chol	10.0000	1.212

Figure 13.17: Odds Ratio for 10 Units of Cholesterol

You see that for every 10 units of cholesterol, the odds ratio is 1.212, approximately a 21% increase in risk (again, inflated from reality for demonstration purposes).

Running a Model with Three Classification Variables

The Risk data set contains three variables that you suspect are related to cardiovascular health. They are: Age_Group, Gender, and Chol_High. The age group variable has three groups: 1:Less 60, 2:61–70, and 3:Over 70. The numbers 1, 2, and 3 in the values for Age_Group will enable you to control the order of the age groups when the analysis is run. Gender is the same as before (F and M), and Chol_High is a dichotomous variable with values of 1 for high and 0 for normal.

Figure 13.18 shows you have selected these three classification variables on the DATA tab.

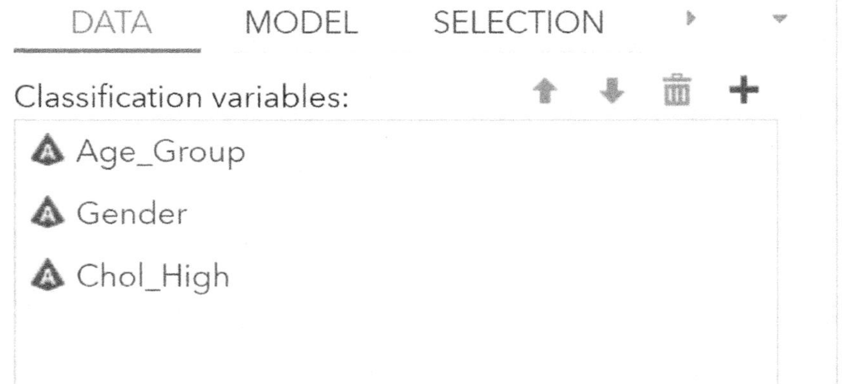

Figure 13.18: Using Three Predictor Variables in Your Model

As in the previous models, you want to select Heart_Attack as your Response variable and a value of 1 as the event of interest. Also, choose Reference coding on the DATA tab in the pull-down list under Parameterization of Effects. Before you run the model, you want to specify reference levels for all three variables. You want to choose the youngest age group for the Age_Group variable, F for Gender, and 0 (normal) for the variable Chol_High. The edited code is displayed in Figure 13.19.

```
proc logistic data=STATS.RISK;
   class Age_Group (ref='1:less 60')
         Gender (ref='F')
         Chol_High (ref='0') / param=ref;
   model Heart_Attack(event='1')=Age_Group Gender
         Chol_High Chol / link=logit
      technique=fisher;
run;
```

Figure 13.19: Editing the Code to Select Reference Levels

It's time to run the model. The fit statistics in Figure 13.20 show a lower value for AIC and SC (in the Intercept and Covariates column), indicating a better model than any of the previous models.

Model Fit Statistics		
Criterion	**Intercept Only**	**Intercept and Covariates**
AIC	360.879	322.045
SC	365.093	343.118
-2 Log L	358.879	312.045

Figure 13.20: Fit Statistics for the Model

Significance tests for all three are highly significant (Figure 13.21).

Type 3 Analysis of Effects			
Effect	**DF**	**Wald Chi-Square**	**Pr > ChiSq**
Age_Group	2	15.4238	0.0004
Gender	1	11.0624	0.0009
Chol_High	1	14.2830	0.0002

Figure 13.21: Probabilities for Each Variable in the Model

The odds ratios (see Figure 13.22) show that there is a very significant increase in risk for older subjects, that men are more likely to have heart attacks compared to women, and that high cholesterol is a significant risk factor.

Odds Ratio Estimates			
Effect	Point Estimate	95% Wald Confidence Limits	
Age_Group 2:61 to 70 vs 1:Less 60	3.878	1.414	10.637
Age_Group 3:Over 70 vs 1:Less 60	7.105	2.615	19.308
Gender M vs F	2.885	1.545	5.387
Chol_High 1 vs 0	3.503	1.828	6.710

Figure 13.22: Odds Ratio Estimates

The display of concordant and discordant pairs (72.2% versus 19.4%) also demonstrates that the model is doing a good job.

Association of Predicted Probabilities and Observed Responses			
Percent Concordant	72.2	Somers' D	0.527
Percent Discordant	19.4	Gamma	0.575
Percent Tied	8.4	Tau-a	0.108
Pairs	25636	c	0.764

Figure 13.23: Concordant and Discordant Pairs

Conclusion

Programming a binary logistic regression model by writing the actual SAS code is a bit daunting. However, by using the Binary Logistic Regression task, it is quite easy to do. This is one instance where even a veteran programmer might decide that a point-and-click approach to running this task is a good choice.

Chapter 13 Exercises

1. The Graduate data set in the STATS library contains information about student graduation rates, the number of hours spent studying each week, along with course grades in English and mathematics. Start out by running some summary statistics on English_Grade and Math_Grade broken down by the variable Study (0=Below Average, 1=Above Average) and Graduate (1=Yes, 0=No).
2. Using the same data set as exercise 1, run the Binary Logistic Regression task with Graduate as the Response variable and 1 as the Event of interest. Use the reference parameterization method, and make Study=0 the reference level. According to these results, does it pay to study?
3. Rerun exercise 2, adding English_Grade and Math_Grade as continuous variables. Did the model improve?
4. Rerun exercise 3, but determine the odds ratio for a 5-point increase in both English and math grades. Remember, you need to edit the task-generated code and add a UNITS statement.
5. Using the Graduate data set in the STATS library, run a stepwise binary logistic regression model with Graduate as the response variable (event = 1) with English_Grade and Math_Grade as continuous variables and Study as a classification variable. (Hint: You will need to click the SELECTION tab and select Stepwise Selection as the method.) You might want to use a reference level of 0 for Study so that the odds ratio will show that you are more likely to graduate if you study versus if you do not study.

14.

Analyzing Categorical Data

Introduction

One of the data analysis tasks that you will be using quite often is to compute frequencies on categorical data and to analyze **two-way frequency tables** (tables with frequencies of one variable making up rows of the table and frequencies of another variable making up columns of the table). You might want to see survey results showing the number and percentage of subjects in one or more categories. Another common statistical test performed on categorical data is to see whether two categorical variables are related. For example, are people with high cholesterol more likely to suffer a heart attack compared to people with normal cholesterol levels? On average, do women earn less than men based on a sample of weekly salaries? This chapter covers two of the SAS Studio statistical tasks that deal with frequency data: the first is called One-Way Frequencies, the other, Table Analysis.

Describing the Salary Data Set

One of the data sets created when you ran the Create_Datasets.sas program is called Salary, and it is placed in the STATS library. This data set contains simulated data on weekly salaries broken down by gender, age group, and educational level. Besides the actual weekly salary (variable Weekly_Salary), there is another numeric variable (Salary) that has values of 0 (salary below the median) and 1 (salary above the median). Although this data set was simulated, the values were based on a data set from the U.S Department of Labor. The Department of Labor data contains salaries for multiple categories: age group, level of education, gender, and several other variables. For simplicity, the Salary data set contains only two levels for most of these variables (typically ones with extreme differences, such as ages 20–24 compared to ages 45–54, where large differences in salaries exist).

Before we get into the actual analysis of this data set, it is important to understand how these statistics are collected. One method, called **uncontrolled**, simply looks at median salaries for each category of predictor, such as gender or age group. The other, called **controlled**, looks at salaries for identical jobs and other factors such as years of employment. All of the data values for this simulated data set are based on the uncontrolled labor statistics values.

Variables in the Salary data set are displayed in the table that follows.

Variable	Description	Values
Weekly_Salary	Weekly salary	Actual amount in dollars
Salary	Above or below the median salary	0 = Below, 1 = Above
Gender	Gender	M = Male, F = Female
Age_Group	Age group	20–24, 45–54
Education	Educational level	Less than HS, College Degree or higher

Computing One-Way Frequencies

To compute **one-way frequencies** (frequencies for a single variable), go to the Statistics tab under Tasks to see a list of statistical tasks (Figure 14.1).

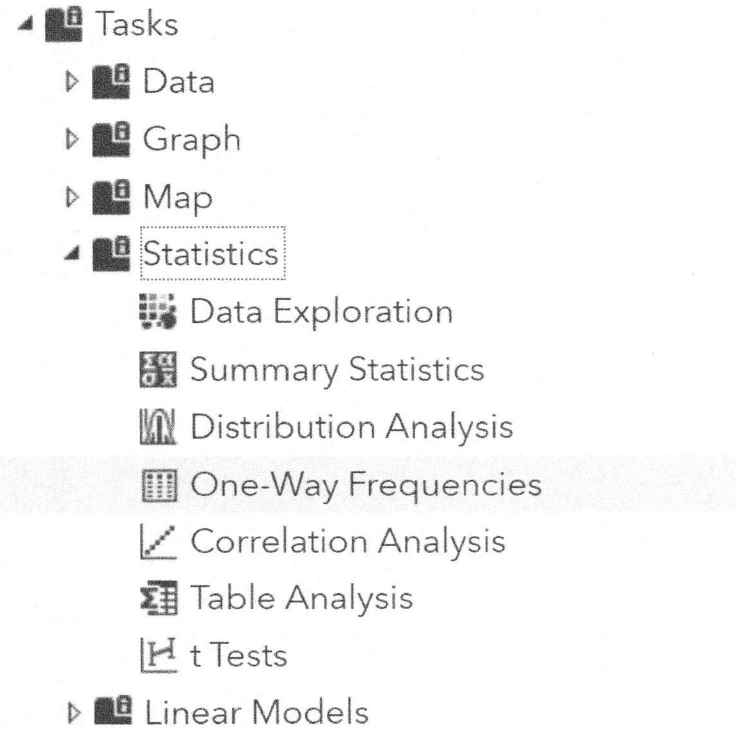

- Tasks
  - Data
  - Graph
  - Map
  - Statistics
    - Data Exploration
    - Summary Statistics
    - Distribution Analysis
    - One-Way Frequencies
    - Correlation Analysis
    - Table Analysis
    - t Tests
  - Linear Models

Figure 14.1: Demonstrating the One-Way Frequency Task

Double-click One-Way Frequencies in the Statistics task list to bring up the following screen (Figure 14.2).

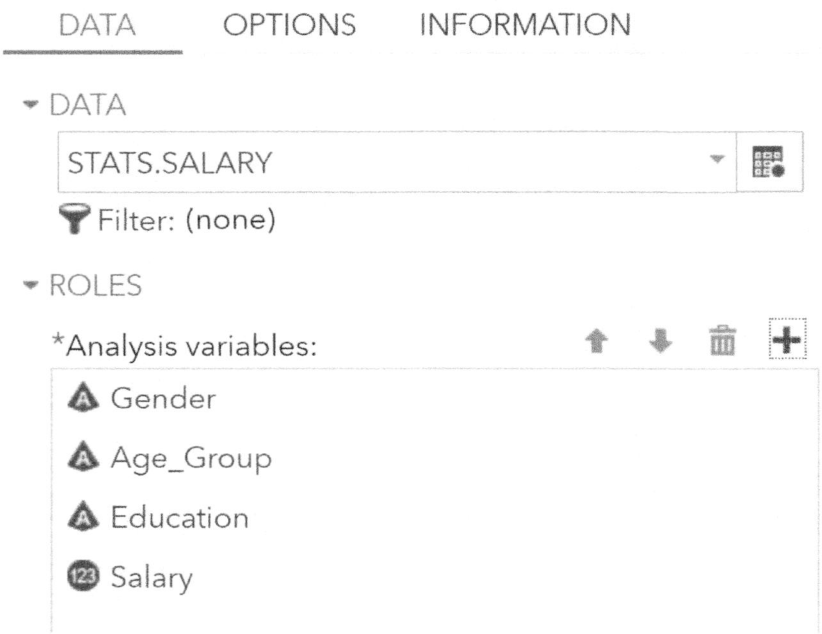

DATA OPTIONS INFORMATION

DATA

STATS.SALARY

Filter: (none)

ROLES

*Analysis variables:

Gender

Age_Group

Education

Salary

Figure 14.2: DATA Tab Selections

On the DATA tab, choose the Salary data set stored in the permanent STATS library. Next, select the variables Salary, Gender, Age_Group, and Education in the Analysis variables box. Notice that the variable Weekly_Salary (which is a numeric variable) is included in your list of choices. If you include numeric variables with many different values (such as Weekly_Salary), the One_Way Frequency task will list frequencies for every unique value of the selected variable. If you included the variable Weekly_Salary as an analysis variable, it would tell you how many people earned $800, how many people earned $801 dollars, and so on. The only numeric variable chosen in this example is Salary. This is OK because this numeric variable is coded as 0 or 1, representing below the median and above the median.

Before you run the procedure, click the OPTIONS tab to select additional options.

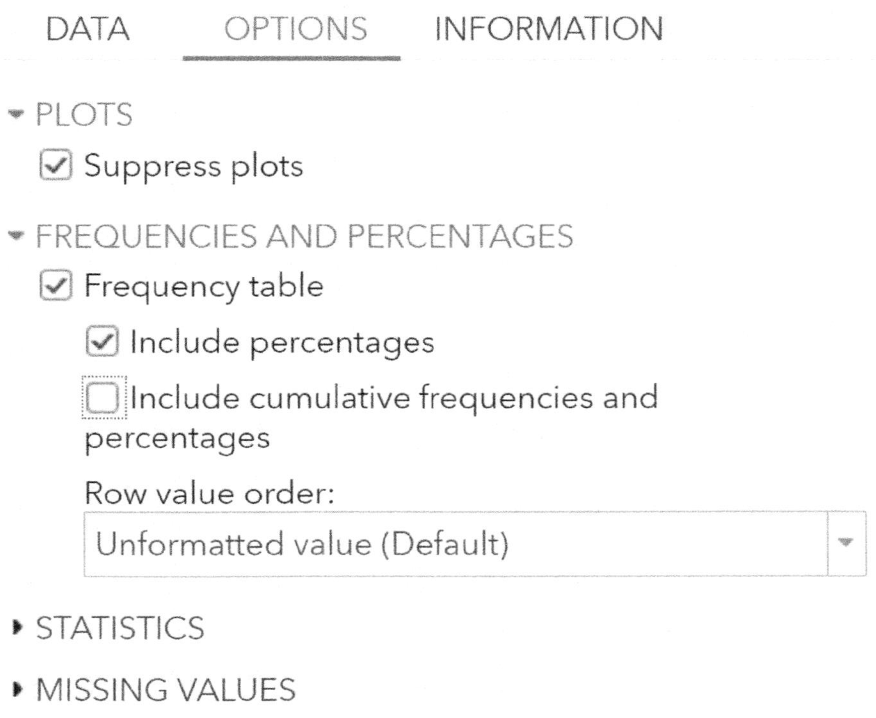

DATA OPTIONS INFORMATION

▾ PLOTS

☑ Suppress plots

▾ FREQUENCIES AND PERCENTAGES

☑ Frequency table

☑ Include percentages

☐ Include cumulative frequencies and percentages

Row value order:

Unformatted value (Default)

▸ STATISTICS

▸ MISSING VALUES

Figure 14.3: One-Way Frequency Options

In this example, you have chosen to suppress plots and to deselect the default option to include cumulative frequencies (which you rarely need). You are now ready to run the procedure. The output is shown in Figure 14.4.

Gender	Frequency	Percent
F	4000	45.45
M	4800	54.55

Age_Group	Frequency	Percent
20-24	4400	50.00
45-54	4400	50.00

Education	Frequency	Percent
<HS	4400	50.00
BA+	4400	50.00

Rank for Variable Weekly_Salary		
Salary	Frequency	Percent
0	4399	49.99
1	4401	50.01

Figure 14.4: Frequency Tables

You see the frequency and percent for each unique value of these variables. Although this is useful information, it could be improved by replacing the values of Gender (F and M) with the labels Male and Female and replacing the values of Salary (0 and 1) with the labels Below the Median and Above the Median. You can define formats to associate each of these unique values with labels. That is the topic of the next section.

Creating Formats

You will need to write a few lines of SAS code to create the formats that you will use to label the values in these tables. You create formats with a SAS procedure called PROC FORMAT. Program 14.1 is a program that creates two formats and then creates a copy of the data set STATS.Salary with a new data set called STATS.Salary_Formatted, also in the STATS library where the new formats are associated with the appropriate variables.

```
*Adding formats to the Salary Dataset; ①

proc format; ②
   value $Gender 'F' = 'Female' ③
                 'M' = 'Male';
   value Median 0 = 'Below the Median' ④
                1 = 'Above the Median';
run;

data Stats.Salary_Formatted; ⑤
   set Stats.Salary; ⑥
   format Gender $Gender. ⑦
          Salary Median.;
run;

title "Listing the First 10 Observations from Stats.Salary_Formatted"; ⑧
proc print data=Stats.Salary_Formatted(obs=10);
run;
```

Program 14.1: Adding Formats to the Program

(1) The first line of the program is a comment statement stating the purpose of the program. Comment statements start with an asterisk and end with a semicolon. Comment statements are optional, but highly recommended.

(2) PROC FORMAT is used to create SAS formats.

(3) Use a VALUE statement to define each format that you want to create. Follow the keyword VALUE with a format name. Format names are a maximum of 32 characters and must contain only letters, digits, and the underscore character. One additional rule is that format names cannot end in a digit. Finally, if the format is to be used with character data (such as Gender), you use a dollar sign as the first character in the format name. (Note: the $ character counts as one of the 32 characters in the format name.) Although the format $Gender is going to be used with the variable called Gender, the format name can be anything you want. It could have been named $Oscar.

Following the format name, you define labels for each value that you want to label. Because Gender is a character variable, the values ‘M’ and ‘F’ must be in single or double quotation marks. Following each value (or a list of values separated by commas), you type an equal sign and the format label. This label also belongs in single or double quotation marks.

(4) The Median format will be used for the variable Salary. Because this variable is numeric, the format name does not start with a $.

You can read more about how to create formats in several SAS Press books, including *Learning SAS by Example, 2nd edition* (Cody, 2018) or *Getting Started with SAS Programming Using SAS Studio in the Cloud* (Cody, 2021).

(5) The DATA statement is creating a new data set called STATS.Salary_Formatted.

(6) The SET statement reads observations from an existing SAS data set. Essentially, this makes a copy of the permanent data set STATS.Salary. The new data set (STATS.Salary_Formatted) will associate the formatted values for the internal values (such as 0 or 1) in the output of many procedures.

(7) The FORMAT statement associates variable names and formats. In this FORMAT statement, the format $Gender is used to format the variable Gender. Notice the period following the format name. This tells the DATA step that $Gender is a format and not a variable name. The format Median is associated with the variable Salary. You end the DATA step with a RUN statement.

(8) PROC PRINT is used to list the first 10 observations in the new data set STATS.Salary_Formatted. The data set option OBS= limits the listing to 10 observations. As an alternative, you could use the List Data task of SAS Studio to list these observations.

The output is displayed in Figure 14.5.

List Data for STATS.SALARY_FORMATTED

Obs	Gender	Age_Group	Education	Weekly_Salary	Salary
1	Female	20-24	<HS	$443	Below the Median
2	Female	20-24	BA+	$1,574	Above the Median
3	Female	45-54	<HS	$1,089	Below the Median
4	Female	45-54	BA+	$1,861	Above the Median
5	Male	20-24	<HS	$939	Below the Median
6	Male	20-24	<HS	$872	Below the Median
7	Male	20-24	<HS	$819	Below the Median
8	Male	20-24	<HS	$873	Below the Median
9	Male	20-24	<HS	$970	Below the Median
10	Male	20-24	<HS	$680	Below the Median

Figure 14.5: First 10 Observations from Data Set STATS.Salary_Formatted (with Formats)

Notice that the formats that you created are displayed in this listing. Because the association between the variables and formats was executed in a DATA step, this association will be maintained in other SAS procedures, such as the one-way or two-way tables.

> It is important to remember that the actual values for the formatted variables still exist in the SAS data set. The formats labels only appear when you use certain procedures that present data, such as the listing above or the tables that you will be producing in this chapter.

Producing One-Way Tables with Formats

If you rerun One-Way frequencies with the new data set Salary_Formatted (the one with associated formats), the one-way frequency tables will now display formatted values. Only two tables (for the variables Gender and Salary) are shown here in Figure 14.6.

Gender	Frequency	Percent
Female	4000	45.45
Male	4800	54.55

Rank for Variable Weekly_Salary		
Salary	Frequency	Percent
Below the Median	4399	49.99
Above the Median	4401	50.01

Figure 14.6: One-Way Frequencies with Formatted Values

These tables have a clear advantage over the unformatted tables produced earlier. It saves you the trouble of going back to your coding scheme to see what Salary 0 or Salary 1 represent. It is pretty clear that F is Female and M is Male, but formatting the values still makes the output easier to read.

Reviewing Relative Risk, Odds Ratios, and Study Designs

In the previous chapter, you learned about odds ratios (abbreviated OR). Here we want to concentrate more on **relative risk** (abbreviated RR and sometimes called risk ratio).

Chapter 2 described several popular study designs. Two of these designs were cohort studies and case-control studies. Cohort studies typically use RRs, while case-control studies typically use ORs to explain their findings. In a cohort study, you have a group of subjects, such as males and females, and you measure something about them (for example, if their salary is below the median level or not). Suppose you have 1,000 women and 1,000 men in a salary study, and you find that 600 of the women had salaries below the median salary and 400 of the men had salaries below the median salary. The proportion of women below the median is 600/1,000 = .6 and the proportion of men below the median salary is 400/1,000 = .4. One way to express this result is to take the ratio of these proportions .6/.4 = 1.5 and claim that the relative risk of having a salary below the median salary for women compared to men is 1.5. Another way to express this is that women are 50% more likely to have salaries below the median salary compared to men.

Let's also review case-control studies. For this study design, you start out by looking at people who have a particular outcome, such as having a rare brain tumor (these people are called cases), and another group of comparable people who do not have brain tumors (these people are called controls). You then look at one or more possible risk factors for this type of tumor, such as exposure to certain toxic chemicals. Let's use benzene as an example. You then compare the proportion of the cases who were exposed to benzene and the proportion of controls who were exposed to benzene. Because this is a case-control study, you cannot compute incidence rates for the brain tumors.

In order to compute incidence rates, you would need a very large group (cohort) of people who were exposed to benzene and a very large group of people who were not exposed to benzene. You would then need to determine the number of rare brain tumors in each group. Furthermore, these tumors might take a long time to grow and be detected, so you might be setting yourself up for a very expensive, long-term study.

Now, back to the case-control study: if you see a higher proportion of cases were exposed to benzene compared to controls, you might expect that there is a relationship between exposure and brain tumors.

Suppose you use a cancer registry to locate 100 people with a particular type of brain tumor and then find 100 people who are comparable to the cases on several factors (gender, age, and so on) but do not have brain tumors. These are the controls. Suppose the results are that 40 cases were exposed to benzene (and 60 were not) and 10 controls were exposed to benzene (and 90 were not). The odds of the cases being exposed is computed as 40 divided by 60, and the odds for the controls being exposed is computed as 10 divided by 90. You can then compute the odds ratio as 40/60 divided by 10/90 = 6.

Because you believe that the incidence rate for this particular form of brain tumor is very rare, you can use the odds ratio as an estimate for the relative risk. (A rule of thumb is that if the incidence rate is less than 10%, the odds ratio is a fairly good estimate of the relative risk.) SAS Studio does not know if your two-way table resulted from a cohort study or a case control study; therefore, it computes both relative risk and odds ratios.

Creating Two-Way Tables

To see relationships between variables, such as if women are more likely to have salaries below the median level compared to men, you want to produce **two-way tables**. The first step is to double-click Table Analysis in the Statistics task menu (Figure 14.7).

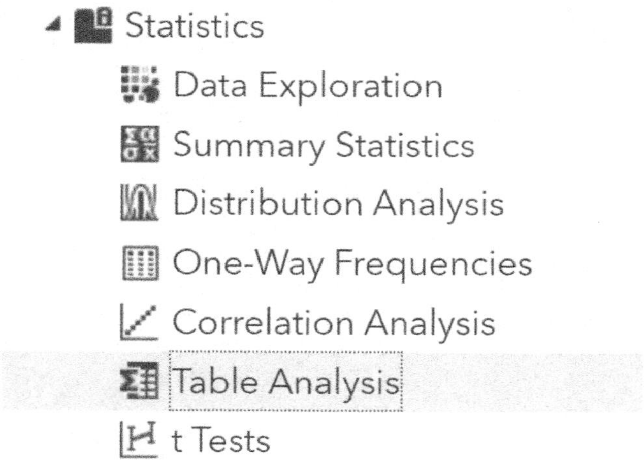

Statistics

Data Exploration

Summary Statistics

Distribution Analysis

One-Way Frequencies

Correlation Analysis

Table Analysis

t Tests

Figure 14.7: Creating a Two-Way Table

Next, fill in the DATA tab.

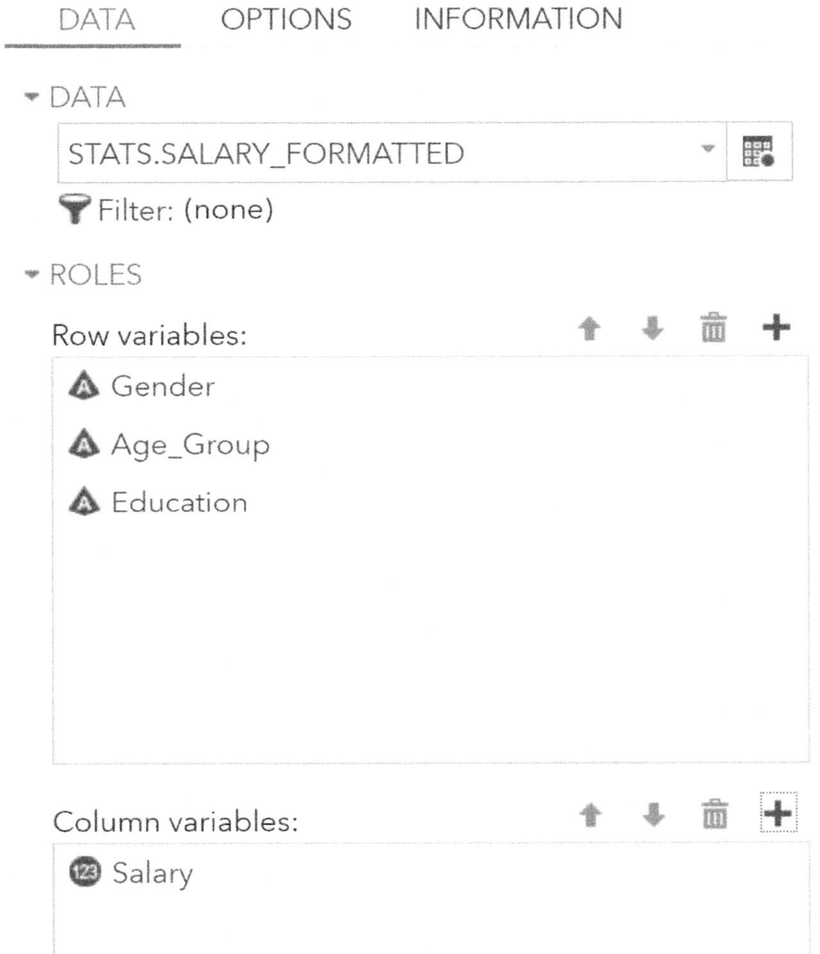

Figure 14.8: Completing the DATA Tab for a Two-Way Table

Notice that the STATS.Salary_Formatted data set (the one with formats) is selected as the data source. Next, you get to choose which variables form the rows of the table and which variables form the columns. Every variable selected in the Row variables box will be paired with every variable in the Column variables box. It is typical to choose the outcome variable (in this case, having a salary below the median level or above the median level) as the column variable and the other variables such as Gender, Age_Group, and Education as row variables.

There are many more options to choose from in the Table task compared to the One-Way task. Click the OPTIONS tab to get started. (See Figure 14.9.)

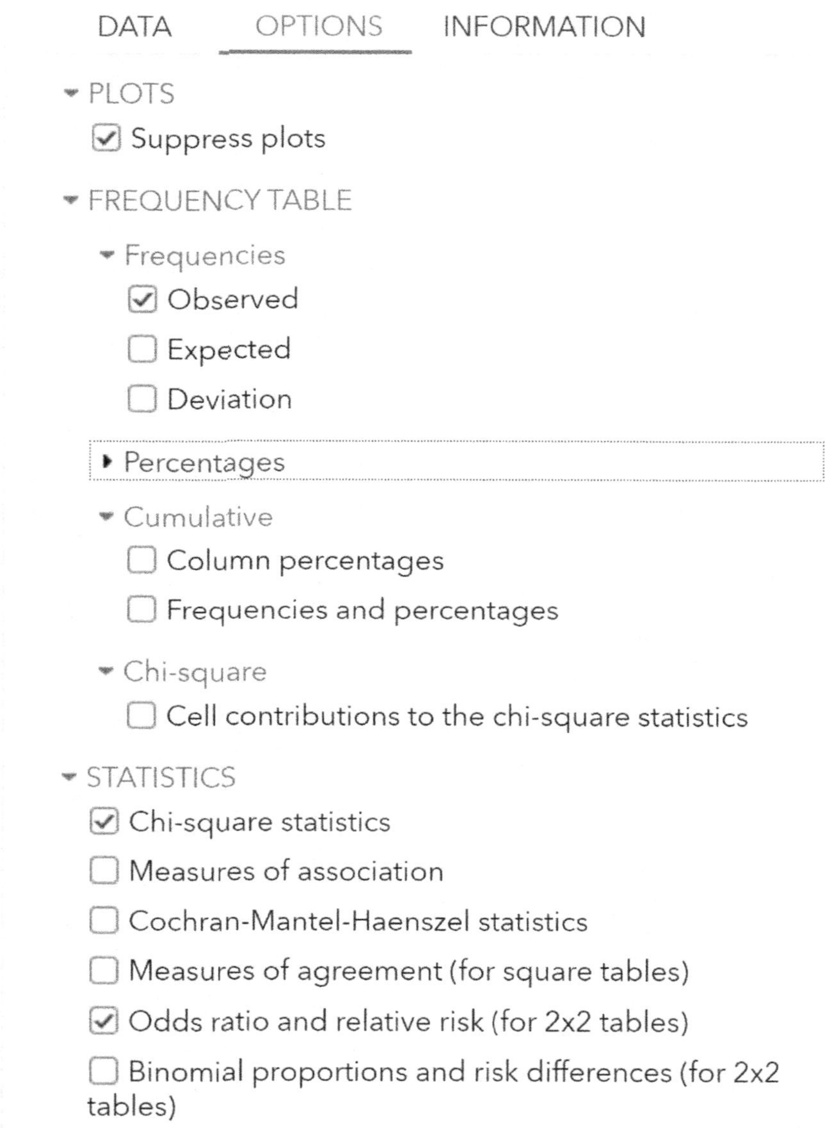

DATA OPTIONS INFORMATION

▾ PLOTS

☑ Suppress plots

▾ FREQUENCY TABLE

▾ Frequencies

☑ Observed

☐ Expected

☐ Deviation

▸ Percentages

▾ Cumulative

☐ Column percentages

☐ Frequencies and percentages

▾ Chi-square

☐ Cell contributions to the chi-square statistics

▾ STATISTICS

☑ Chi-square statistics

☐ Measures of association

☐ Cochran-Mantel-Haenszel statistics

☐ Measures of agreement (for square tables)

☑ Odds ratio and relative risk (for 2x2 tables)

☐ Binomial proportions and risk differences (for 2x2 tables)

Figure 14.9: Options for a Two-Way Table

Check Suppress plots if you do not want plots. Next, you have a choice of what percentages you would like to display. Here, you have chosen to see row and column percentages. Because you don't want to see cumulative frequencies (not usually useful), make sure that this box is deselected.

As you can see, there are many options in the statistics menu. Two of the more common options, Chi-square and Odds ratios and relative risks, were chosen for this example. It's time to run the task. Here are the results. (Note: to save space, only one table was included here.)

Frequency Row Pct Col Pct

Table of Gender by Salary			
	Salary(Rank for Variable Weekly_Salary)		
Gender	Below the Median	Above the Median	Total
Female	2310 57.75 52.51	1690 42.25 38.40	4000
Male	2089 43.52 47.49	2711 56.48 61.60	4800
Total	4399	4401	8800

Figure 14.10: Table of Gender by Median Cut

The box in the upper left corner of the output is the key to the three numbers in each box. The top number is a frequency count. For example, there were 2,310 workers who were female and had a weekly salary below the median salary. The second number in each cell is a row percentage. In this example, 57.75% of the workers who were female had salaries below the median salary. The third number in each box is a column percentage. In this example, 52.51% of the workers who had salaries below the median were female.

The next portion of the output offers you several statistical tests for the association of gender with median salary.

Statistics for Table of Gender by Salary

Statistic	DF	Value	Prob
Chi-Square	1	176.7004	<.0001
Likelihood Ratio Chi-Square	1	177.3160	<.0001
Continuity Adj. Chi-Square	1	176.1317	<.0001
Mantel-Haenszel Chi-Square	1	176.6803	<.0001
Phi Coefficient		0.1417	
Contingency Coefficient		0.1403	
Cramer's V		0.1417	

Figure 14.11: Measures of Association

The most commonly used statistic when you have a large number of subjects is **chi-square**. This statistic compares observed frequencies to frequencies that you would expect if the null hypothesis were true (called expected frequencies). The relationship, by this measure, is highly significant ($p < .0001$). Some of the other statistics in this table, such as Continuity Adj. Chi-Square, are useful when you have small expected values for any of the table frequencies. One of the assumptions for computing chi-square is that none of the expected frequencies in a 2-by-2 table is less than 5. Dealing with violations of this assumption is addressed later in this chapter.

The final portion of the output for the variables Gender and Salary was printed because you checked the box labeled Odds Ratios and Relative Risk on the OPTIONS tab. It is shown next.

Odds Ratio and Relative Risks			
Statistic	**Value**	**95% Confidence Limits**	
Odds Ratio	1.7738	1.6296	1.9308
Relative Risk (Column 1)	1.3270	1.2727	1.3835
Relative Risk (Column 2)	0.7481	0.7159	0.7817

Figure 14.12: Odds Ratios and Relative Risk

The number that you should focus on is the Relative Risk (Column 1) equal to 1.3270, because this is a cohort study and relative risk is usually the statistic of choice for this type of study. The value of 1.3270 is the incidence of having a weekly salary below the median if you are female compared to the weekly salary if you are male.

Using Formats to Reorder the Rows and Columns of a Table

By chance, the rows and columns in the above table were in an order that made sense: You wanted to know if being female makes it more likely that your salary would be below the median salary.

The ordering of the rows and columns in the Table Analysis task are, by default, arranged by the internal ordering of the values used in the table. Females form the first row of the table because the internal value of F comes alphabetically before M. The columns are ordered with Below the Median as the first column and Above the Median as the second column because the internal values were 0 and 1 respectively. There are times when it makes sense to change the order of the rows and/or columns.

To demonstrate this, let's look at an example from the Risk data set where it is clear that you would like to change the order of the rows and columns. First, let's look at a table of having a heart attack versus cholesterol level.

Frequency Row Pct Col Pct

Table of Chol_High by Heart_Attack			
	Heart_Attack		
Chol_High	0	1	Total
0	231 94.29 52.26	14 5.71 24.14	245
1	211 82.75 47.74	44 17.25 75.86	255
Total	442	58	500

Figure 14.13: Table of Heart Attack versus Cholesterol Level

In this table, the variable Heart_Attack is a numeric variable with values of 0 (did not have a heart attack) versus 1 (did have a heart attack). The values for the variable Chol_High are character values of '0' (below 240) and '1' (above 240). The odds ratio and relative risk for this table is shown in Figure 14.14.

Odds Ratio and Relative Risks			
Statistic	Value	95% Confidence Limits	
Odds Ratio	3.4408	1.8331	6.4585
Relative Risk (Column 1)	1.1395	1.0689	1.2147
Relative Risk (Column 2)	0.3312	0.1863	0.5887

Figure 14.14: Odds Ratios and Relative Risk for Heart Attack and Cholesterol Levels

This table shows you the odds ratio and relative risk for **not** having a heart attack if you have **low** cholesterol. It would make more sense to ask for the odds ratio or relative risk for **having** a heart attack if you have **high** cholesterol. To do this, you need to reorder the rows and columns of this table.

There is an option in PROC FREQ called ORDER= that enables you to select several ways to order values in a table. The default order is by the internal values. That is why the table in Figure 14.13 is ordered the way it is. You can set an option called ORDER=FORMATTED with PROC FREQ (the procedure used by the Table Analysis task) to use formatted values rather than the non-formatted (internal) values.

You need to edit the SAS code produced by the Tables Analysis task to create formats to associate these formats with the variables in your table and to add the option ORDER= to PROC FREQ. You start by first clicking the CODE tab and then clicking the Edit button, as shown in Figure 14.15.

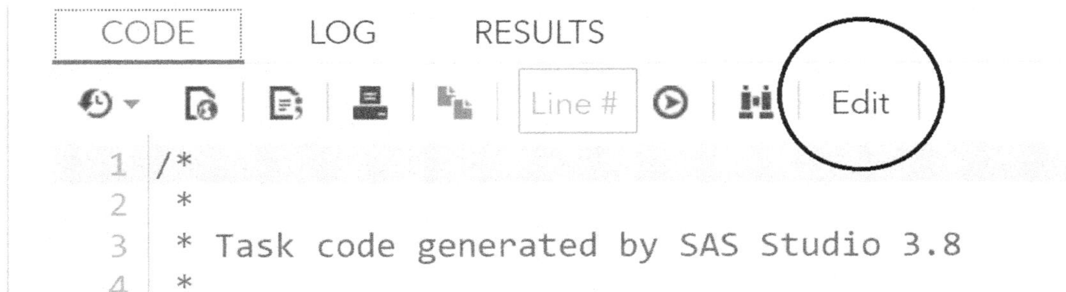

Figure 14.15: Editing the Task-Generated Code

Now, make changes (shown in bold) to the code as pictured in Program 14.2.

```
Proc Format; ①
   value Yesno 1='1:Yes' 0='2:No'; ②
   value $Level '1'='1:High' '0'='Normal'; ③
run;

proc freq data=STATS.RISK order=formatted; ④
   format Heart_Attack Yesno. ⑤
          Chol_High $Level.;
   tables  (Chol_High) *(Heart_Attack) / chisq relrisk
      nopercent norow nocol
      nocum plots=none;
run;
```

Program 14.2: Demonstrating Order=Formatted

(1) Use PROC FORMAT to create your two formats.

(2) Because the variable Heart_Attack is numeric, you need to make a numeric format for this variable. You use the “trick” of placing 1: and 2: in the format labels to force the order to be in the desired direction (Yes before No).

(3) The variable Chol_High is a character variable, so the format name must begin with a dollar sign ($). You use the same trick of using 1: and 2: to force the order.

(4) Add the option ORDER=formatted to PROC FREQ.

(5) Add a FORMAT statement to PROC FREQ to associate the formats with the variables.

With these changes, you can run the program again to obtain the table below.

Frequency Row Pct Col Pct

Table of Chol_High by Heart_Attack			
	Heart_Attack		
Chol_High	1:Yes	2:No	Total
1:High	44 17.25 75.86	211 82.75 47.74	255
2:Normal	14 5.71 24.14	231 94.29 52.26	245
Total	58	442	500

Figure 14.16: Table from Edited Program

The odds ratios and relative risks are now easier to interpret. (See Figure 14.17.)

Odds Ratio and Relative Risks			
Statistic	Value	95% Confidence Limits	
Odds Ratio	3.4408	1.8331	6.4585
Relative Risk (Column 1)	3.0196	1.6987	5.3678
Relative Risk (Column 2)	0.8776	0.8232	0.9356

Figure 14.17: Odds Ratios and Relative Risks from the Modified Program

The relative risk of having a heart attack based on high levels of cholesterol is 3.0196.

Computing Chi-Square from Frequency Data

When you already have a 2-by-2 table with frequency counts and you want to compute chi-square and related statistics, you can still use the Table Analysis task. Suppose you have a table with the data shown in Figure 14.18. You can put the data from this table in an Excel worksheet as follows.

	A	B	C
1	Row	Column	Count
2	1	1	44
3	1	2	211
4	2	1	14
5	2	2	231
6			

Figure 14.18: Spreadsheet Containing Counts

The next step is to convert this data into a SAS data set using the Import utility. Figure 14.19 is a listing of the SAS data set created from the frequency data. The WORK data set was called Counts.

List Data for WORK.COUNTS

Obs	Row	Column	Count
1	1	1	44
2	1	2	211
3	2	1	14
4	2	2	231

Figure 14.19: Data Imported into a SAS Data Set

Start the Table Analysis task and use the DATA tab to identify the variable Row as the row variable and the variable Col as the column variable.

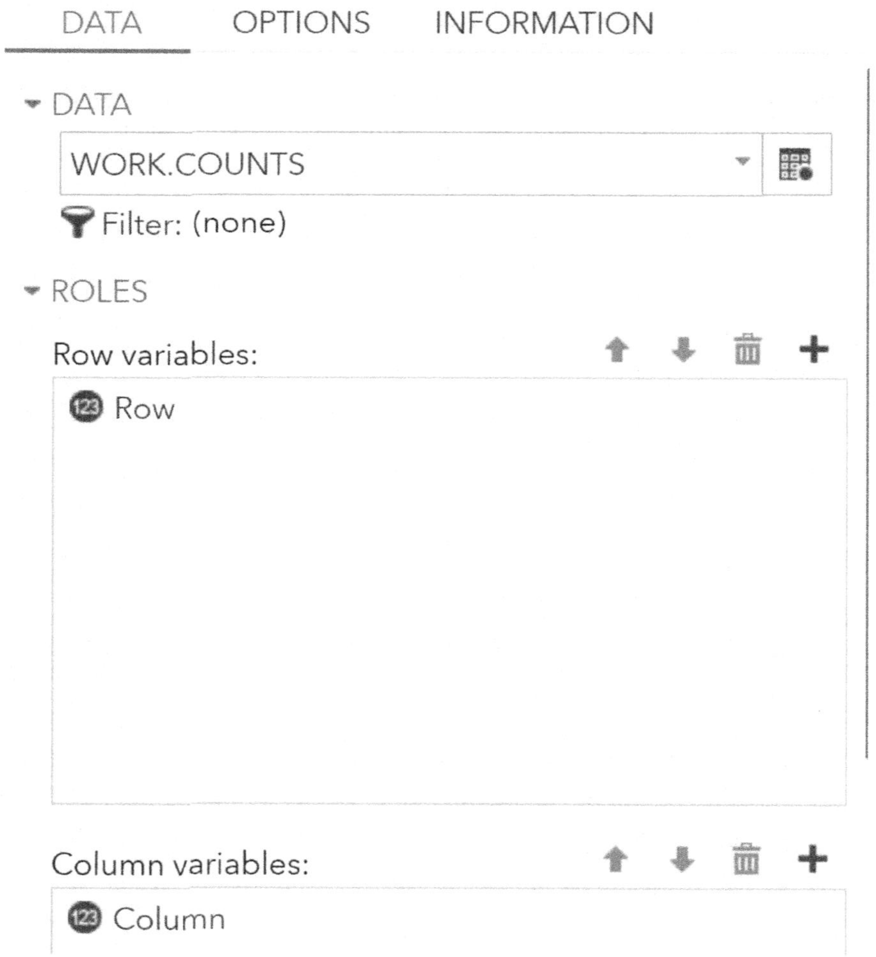

DATA OPTIONS INFORMATION

DATA

WORK.COUNTS

Filter: (none)

ROLES

Row variables:

Row

Column variables:

Column

Figure 14.20: Table Analysis Data Tab

In order to tell the task that you already have frequency data rather than raw data, click Additional Roles at the bottom of the Data screen (Figure 14.21).

Figure 14.21: Identifying the Variable Representing Frequencies

Click the plus sign and select the variable Count as the variable that represents the frequency data. Finally, select the OPTIONS tab and select the statistics that you want. The 2-by-2 table (shown below) now shows the value of the variable Count in the appropriate boxes.

Frequency Row Pct Col Pct

Table of Row by Column			
	Column(Column)		
Row(Row)	**1**	**2**	**Total**
1	44 17.25 75.86	211 82.75 47.74	255
2	14 5.71 24.14	231 94.29 52.26	245
Total	58	442	500

Figure 14.22: Resulting Table

Analyzing Tables with Low Expected Values

The final topic in this chapter deals with 2-by-2 tables with expected values less than 5. As mentioned earlier in this chapter, one of the assumptions for computing chi-square in a 2-by-2 table is that none of the expected values are less than 5. When this assumption is not met, a popular alternative to chi-square is **Fisher's Exact test**.

There are statisticians who prefer chi-square with a correction for continuity instead of Fisher's Exact test. SAS produces both statistics, so you can take your pick.

Let's use the table displayed in Figure 14.23 that has at least one expected frequency less than 5, to demonstrate how to analyze tables with small expected values.

Frequency

Table of Row by Column			
	Column		
Row	1	2	Total
1	2	10	12
2	7	4	11
Total	9	14	23

Figure 14.23: Table with Expected Values Less than 5

Remember that the 2-by-2 frequency table displayed in Figure 14.23 contains actual counts, not expected values. You can easily compute the expected value for each cell or let SAS do it. Let's do the latter. First click the OPTIONS tab. Then under Frequency Table, click Frequencies and check the box to display expected values. This adds expected values to each cell. The resulting table looks like this:

Frequency
Expected

Table of Row by Column			
	Column		
Row	1	2	Total
1	2 4.6957	10 7.3043	12
2	7 4.3043	4 6.6957	11
Total	9	14	23

Figure 14.24: Displaying Expected Values for Each Cell

The bottom number in each cell is the expected value for that cell. Notice that two cells have expected values less than 5. To perform Fisher's Exact test, click Exact Test on the OPTIONS tab and check the box for Fisher's Exact test (Figure 14.25). This test is a popular choice for tables with expected values less than 5.

Figure 14.25: Selecting Fisher's Exact Test

It's time to run the task. Figure 14.26 and Figure 14.27 show values for the continuity adjusted chi-square (3.5267, p = .0604) and Fisher's Exact test (two-sided value p = .0361).

Statistics for Table of Row by Column

Statistic	DF	Value	Prob
Chi-Square	1	5.3158	0.0211
Likelihood Ratio Chi-Square	1	5.5550	0.0184
Continuity Adj. Chi-Square	1	3.5267	0.0604
Mantel-Haenszel Chi-Square	1	5.0847	0.0241
Phi Coefficient		-0.4808	
Contingency Coefficient		0.4333	
Cramer's V		-0.4808	
WARNING: 50% of the cells have expected counts less than 5. Chi-Square may not be a valid test.			

Figure 14.26: Chi-Square and Adjusted Chi-Square

Fisher's Exact Test	
Cell (1,1) Frequency (F)	2
Left-sided Pr <= F	0.0291
Right-sided Pr >= F	0.9975
Table Probability (P)	0.0267
Two-sided Pr <= P	0.0361

Figure 14.27: Fisher's Exact Test

Conclusion

Analysis of frequency data is one of the major tools in analyzing outcomes where the recorded values are dichotomous or categorical. The two tasks, One-Way Frequencies and Table Analysis, provide you with all the tools you need. In addition, this chapter showed you how to create formats and to use formats to control the order of values in one-way or two-way tables.

Chapter 14 Exercises

1. Using the data set Graduate in the STATS library, compute one-way frequencies for the variables Gender, Study, and Graduate. Omit cumulative statistics and suppress all plots.
2. Using the data set Graduate in the STATS library, produce a table with Graduate on the column dimension and Study on the row dimension. Request the chi-square statistic and the relative risk.
3. Repeat exercise 2, but use formats to rearrange the columns so that Yes comes before No. Also rearrange the rows so that the top row is Above Average and the second row is Below Average. (Note that the values of Graduate are already in the correct order.) The relative risk (Col 1) is now the "risk" of graduating if you study hard. (Hints: You need to write a format for the variable Graduate. You also need to add the ORDER=formatted option on PROC FREQ.)
   Here is an example of what your table should look like:

Table of Study by Graduate

	Graduate		
Study	1:Yes	2:No	Total
Above Averge	158	5	163
Below Average	96	41	137
Total	254	46	300

4. Using the data set Cars in the SASHELP library, create a two-way table with the variable Cylinders on the column dimension and Type on the row dimension. Omit all plots and statistics. After you run the task, create a filter to limit the number of cylinders to 4, 6, or 8.
5. Using the data set Fish in the SASHELP library, generate a frequency list for the variable Species in decreasing order of frequency. The PROC FREQ option that you need to include is ORDER=Freq.

15.

Computing Power and Sample Size

Introduction

In my 26 years as a biostatistician at the Rutgers Robert Wood Johnson Medical School in New Jersey, the question I was most frequently asked was, "I'm doing an experiment; how many subjects do I need?" Just about any study, especially one for which you are asking for funding or applying for a grant, will require detailed power and sample size calculations.

As discussed in Chapter 1, the **power** of a study is the probability that the study will result in a statistically significant finding if the drug or treatment that you are studying is different (hopefully better) than either a placebo or an alternate drug or treatment. For many studies conducted at research labs or universities, powers of 80% or 90% are typical. Very large-scale studies may strive for a power of 95%. Small, exploratory studies might be satisfied with powers closer to 70%. The bottom line is that it is unethical and wasteful to conduct a study with low power. You will have a low probability of demonstrating the superiority of your drug or intervention, and this negative result may dissuade others from investigating the same drug or intervention, when it may actually be beneficial. You are also possibly putting people at risk when there is a small probability that the study will achieve a positive result.

Depending on the type of study (comparing means or comparing proportions, for example), there is a set of questions that needs to be answered before you can determine the number of subjects you will need for a particular study. SAS Studio includes a Power and Sample Size task that performs this analysis for many of the popular study designs.

Computing Sample Size for a *t* Test

Let's start with a simple study to compare blood pressure in a group of borderline hypertensive subjects. You want to see whether a low dose of a beta blocker will reduce blood pressure. Because these subjects are borderline hypertensive and the trial will be relatively short, you decide that it is ethical to use a placebo as your control.

What information do you need to decide how many subjects you need to recruit in order to have a power of 80%? Let's open up the Power and Sample Size tab (Tasks and Utilities ⇒ Tasks ⇒ Power and Sample Size). It looks like this:

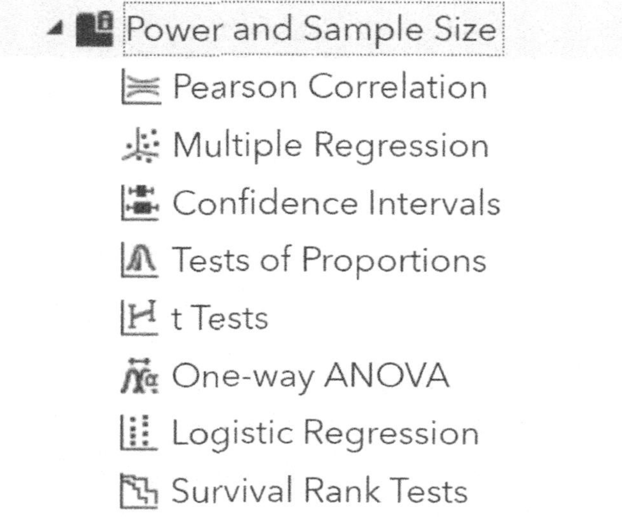

Figure 15.1: Power and Sample Size Menu

Use the t Tests selection to begin the sample size calculation for comparing two means. Double-clicking t Tests brings up the following:

Figure 15.2: Select Type of Test and Sample Size or Power

The menu under Type of *t* test allows you to select from a one-sample test or a two-sample test (either paired or unpaired). Because your study design called for two unpaired groups, select the two-sample option. Next, you can choose to compute power (for a given sample size) or sample size (for a given power). In most studies, you want to know the sample size for a given power. You can request the total sample size or the sample size per group. It is the latter choice that is shown in Figure 15.2.

The next section of the Properties tab asks you to decide if the test is one- or two-sided. Most studies of this type are two-sided. You can also decide if you want to assume equal variances in the two groups. Figure 15.3 shows selections for a two-sided test with equal (pooled) variances.

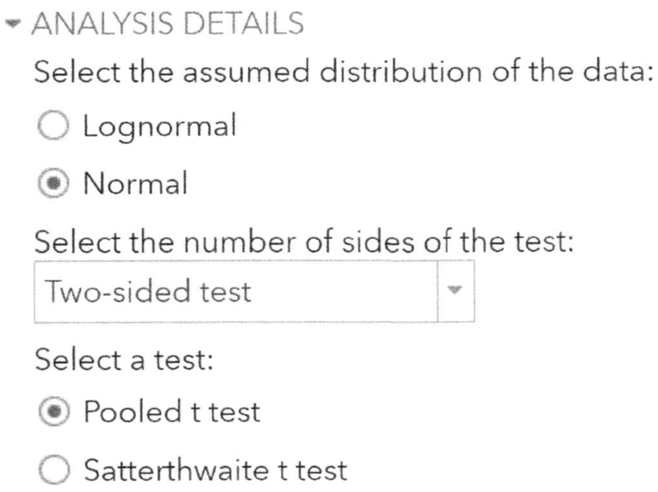

Figure 15.3: Select Distribution, 1- or 2-Tailed, and Variance Assumption

The calculation for power requires you to either estimate the means of the two groups or the difference between the two means. The screen shot in Figure 15.4 shows the different ways that you can enter this information.

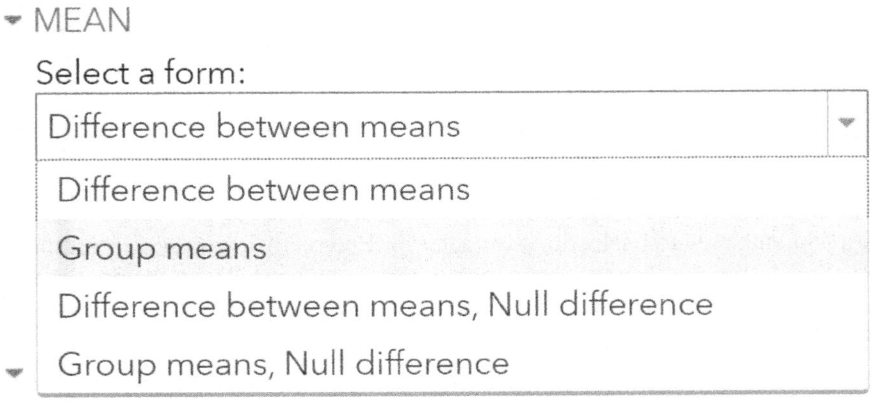

Figure 15.4: Select Ways to Specify Means or Differences

For this example, you are choosing to enter the group means you expect. It is pretty easy to guess the mean for the placebo group because you can take blood pressures on subjects similar to the ones that you will recruit for your study. Choosing one or more means for the treatment group is a bit trickier. Many researchers select values such as a 25% change in the treatment group or a value that they hope to obtain with the treatment, or that is clinically meaningful. With this selection, the menu system opens boxes for you to enter the expected mean for each group (See Figure 15.5).

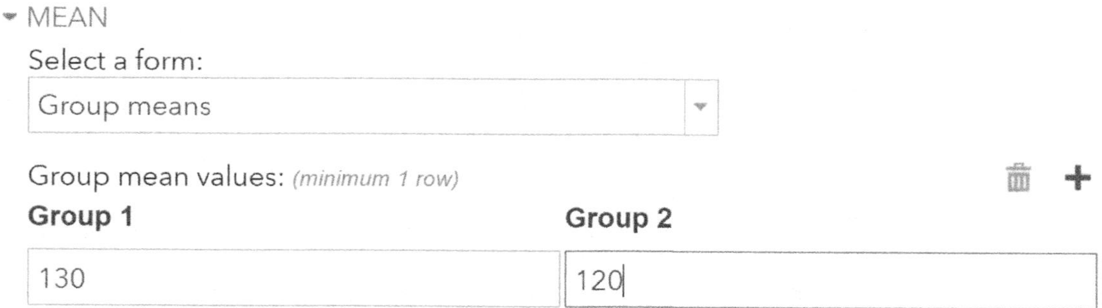

Figure 15.5: Entering Means for Each Group

By clicking the plus sign above the location where you enter your choices for means, you can calculate sample size for different choices of sample means. Figure 15.6 shows that you want to compute sample sizes for means of 130 versus 120 (a 10-point difference) and 130 versus 125 (a 5-point difference).

Group mean values: *(minimum 1 row)*

Group 1	Group 2
130	120
130	125

Figure 15.6: Entering Another Pair of Means

Your final two decisions are to estimate the standard deviation and the desired power. As with other choices in this task, you can enter several selections for each. In Figure 15.7, you see an estimate of 10 for the standard deviation (estimated from measurements from people similar to the ones that you will recruit for your study) and two powers: .8 and .9. Note that powers are entered as probabilities (values between 0 and 1) and not as percentages.

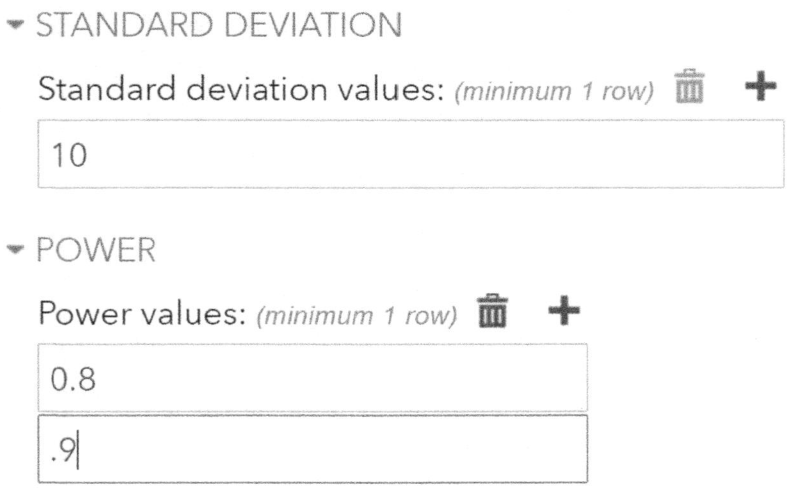

STANDARD DEVIATION

Standard deviation values: *(minimum 1 row)*

10

POWER

Power values: *(minimum 1 row)*

0.8

.9

Figure 15.7: Enter the Desired Power

Before you run the task, click the PLOTS tab and make sure that the box Power by sample size plot is checked. You can let the task scale the power axis, or you can specify minimum and maximum powers by checking the two boxes under Range of values (Figure 15.8).

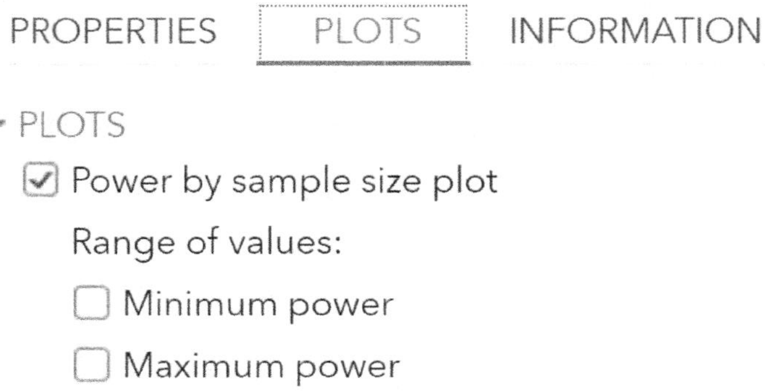

PROPERTIES PLOTS INFORMATION

PLOTS

Power by sample size plot

Range of values:

Minimum power

Maximum power

Figure 15.8: Requesting a Plot of Power by Sample Size

You are ready to run the task. Figure 15.9 shows the results in table form, and Figure 15.10 shows the results in graphical form.

Computed N per Group					
Index	Mean1	Mean2	Nominal Power	Actual Power	N per Group
1	130	120	0.8	0.807	17
2	130	120	0.9	0.912	23
3	130	125	0.8	0.801	64
4	130	125	0.9	0.903	86

Two-Sample t Test for Mean Difference

Figure 15.9: Sample Size per Group in Table Form

Notice the large sample sizes necessary to detect a small difference of 5 points and high power (n = 86 per group). The lowest sample size per group (n per group = 17) is for the largest difference (130 versus 120) and the lowest power (80%).

The graph of sample size by power (see Figure 15.10) contains a line for each combination of means, standard deviations, and powers that you entered. The final decision of sample size is sometimes a compromise between how many subjects you can recruit (and pay for) and how large a difference you would like to be able to detect.

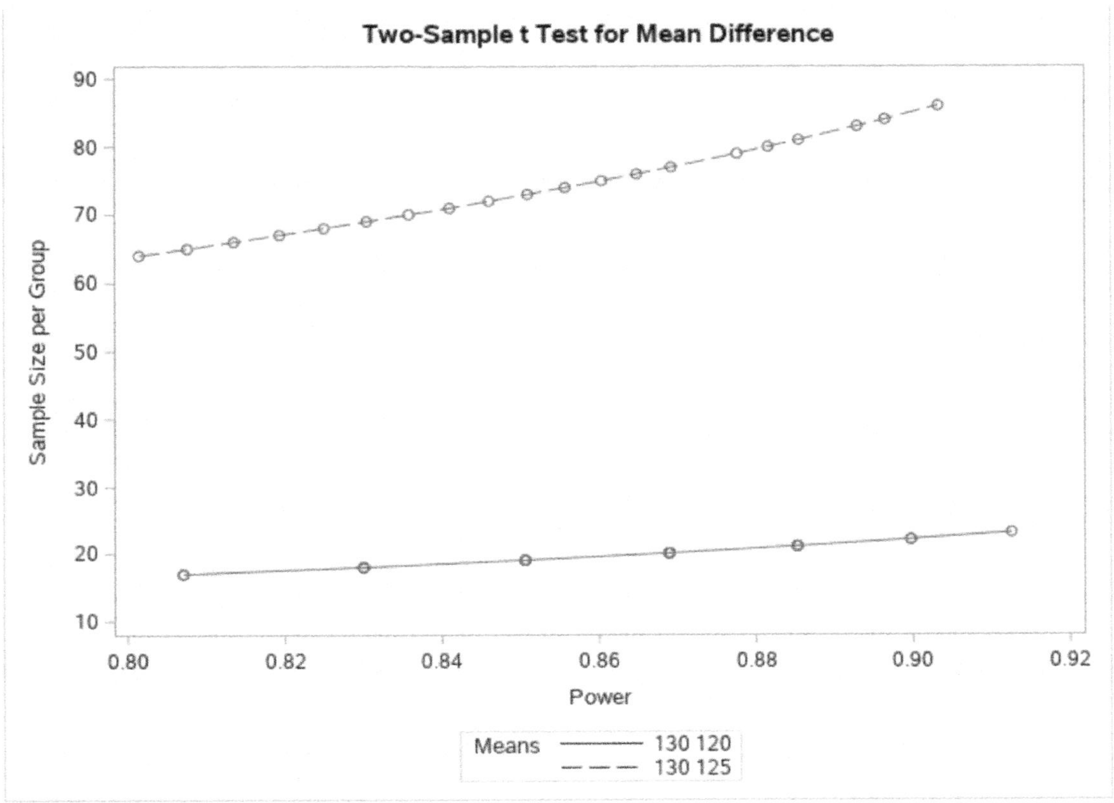

Figure 15.10: Plot of Power versus Sample Size

Calculating the Sample Size for a Test of Proportions

What information do you need to compute sample sizes for a test to compute the difference between two proportions? The decisions for this calculation are simpler than the decisions that you made for comparing means. The reason is that once you select proportions (or difference between proportions), the standard deviation can be computed.

The information that you need to perform this calculation is listed below:

- Is the test to be conducted as a one-sided or two-sided test? (Usually two-sided)
- What is your alpha level? (Usually $\alpha = .05$)
- What is the proportion in the first group (usually a control group)? If unsure, lean toward .5 (maximum variance).

- How large a difference in proportions do you want to be able to detect? (Or the proportion in group two.)
- What power do you want? (You often enter several values such as .8, .85, and .9)

You are now ready to run the Test of Proportions task. Double-click this selection in the Power and Sample Size menu to bring up the following screen. (As before, the Properties screen is shown in pieces.)

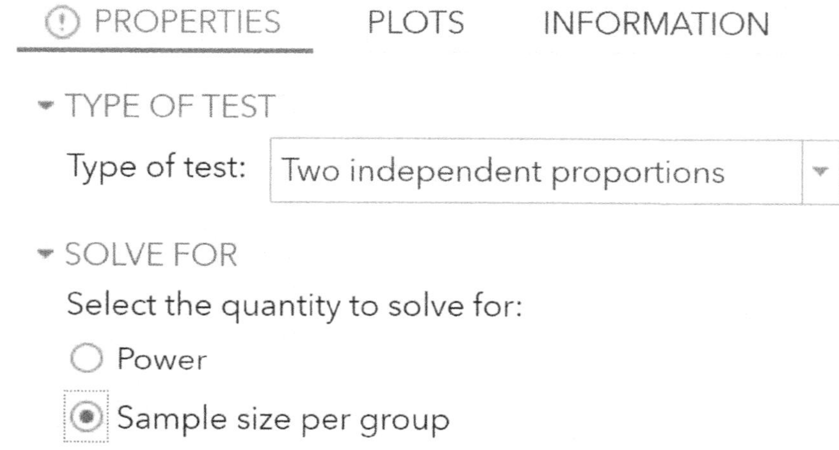

Figure 15.11: Select Type of Test and Solve for Power or Sample Size

For the selection Type of test, select Two independent proportions. Next, decide if you want to test for power or sample size. In this example, you want to know how many subjects you need to obtain a desired power. You have a choice of calculating total sample size or sample size per group. Here, you chose sample size per group.

The next part of the Properties screen asks you to select the statistical test that you plan to use in the analysis. For most studies, especially those with fairly large n's, the Pearson chi-square test is a good choice. If you believe that you will have small expected values in the study, you might choose Fisher's exact test. Finally, indicate if the test is to be conducted as a one-sided or two-sided test. In Figure 15.12, you see selections for a Pearson chi-square test conducted as a two-sided test.

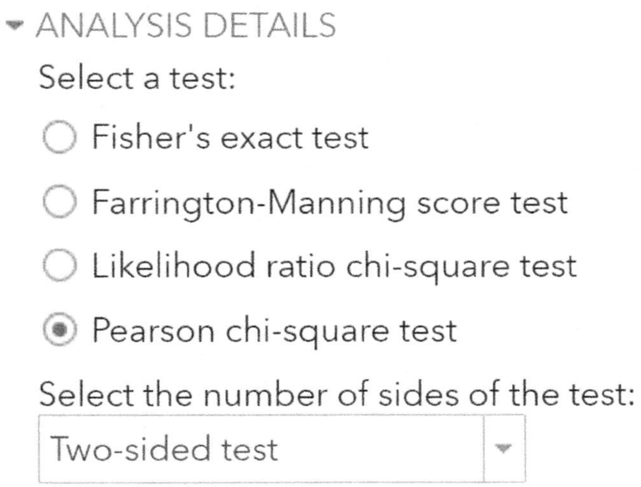

▾ ANALYSIS DETAILS

Select a test:

○ Fisher's exact test

○ Farrington-Manning score test

○ Likelihood ratio chi-square test

◉ Pearson chi-square test

Select the number of sides of the test:

Two-sided test ▾

Figure 15.12: Choose the Statistical Method That You Will Be Using

Moving farther down the menu, you see a choice for alpha (the default value of .05 is already entered) and whether you want to enter group proportions or other ways of comparing proportions such as odds ratios or relative risk. The choices in Figure 15.13 are alpha = .05 and Group proportions. As with the previous calculation where you were comparing means, you can also enter several choices for proportions in the two groups. You decide to compute sample sizes for two different scenarios: one with proportions of .7 and .8, the other with proportions of .7 and .9.

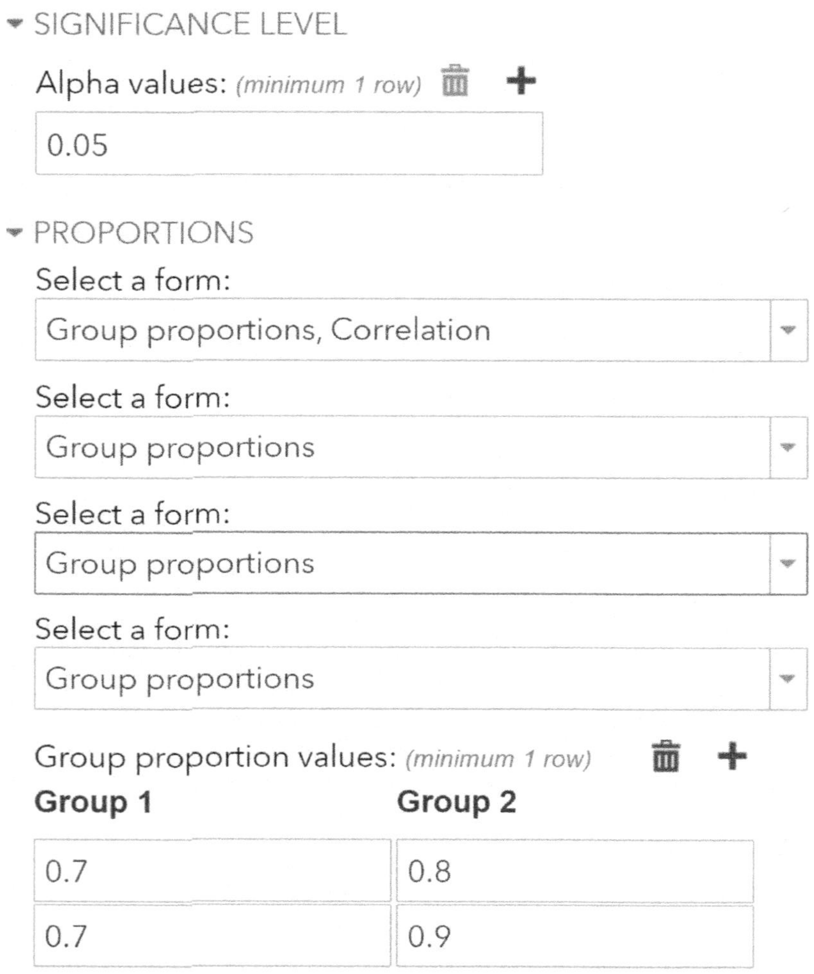

SIGNIFICANCE LEVEL

Alpha values: *(minimum 1 row)*

0.05

PROPORTIONS

Select a form:

Group proportions, Correlation

Select a form:

Group proportions

Select a form:

Group proportions

Select a form:

Group proportions

Group proportion values: *(minimum 1 row)*

Group 1	Group 2
0.7	0.8
0.7	0.9

Figure 15.13: Select Alpha Level and Proportions in the Two Groups

The last entry in the PROPERTIES tab allows you to enter one or more values for power. In Figure 15.14, powers of .8, .85, and .9 were selected.

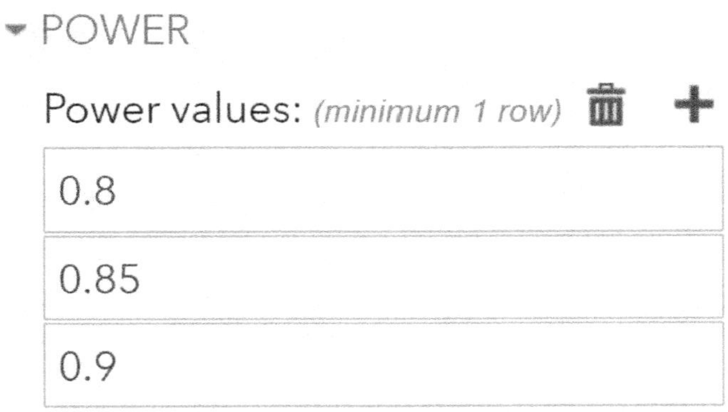

▾ POWER

Power values: *(minimum 1 row)*

0.8
0.85
0.9

Figure 15.14: Select One or More Power Values

Before you run the task, click the PLOTS tab to request a plot of power by sample size.

PROPERTIES PLOTS INFORMATION

▾ PLOTS

☑ Power by sample size plot

Range of values:

☐ Minimum power

☐ Maximum power

Figure 15.15: Select Plot

It's time to run the task. Click the Run icon to obtain the table and graph displayed in Figure 15.16 and Figure 15.17.

Many researchers are shocked when they see the large sample sizes needed to compare proportions. If you look at the N per Group for comparing proportions of .7 and .8 with a power of 90%, you see that you need 392 subjects per group. The smallest number of subjects per group (62) is for proportions of .7 versus .9 with 80% power.

Computed N per Group					
Index	Proportion1	Proportion2	Nominal Power	Actual Power	N per Group
1	0.7	0.8	0.80	0.801	294
2	0.7	0.8	0.85	0.851	336
3	0.7	0.8	0.90	0.900	392
4	0.7	0.9	0.80	0.803	62
5	0.7	0.9	0.85	0.854	71
6	0.7	0.9	0.90	0.900	82

Figure 15.16: Table of Results

You might find it more instructive to examine the sample size calculations in graphical form. As with the previous situation where you were comparing means, when you compare proportions, you may have to lower your expectations of detecting small differences and design the study with larger differences in the two proportions and, perhaps, slightly lower power.

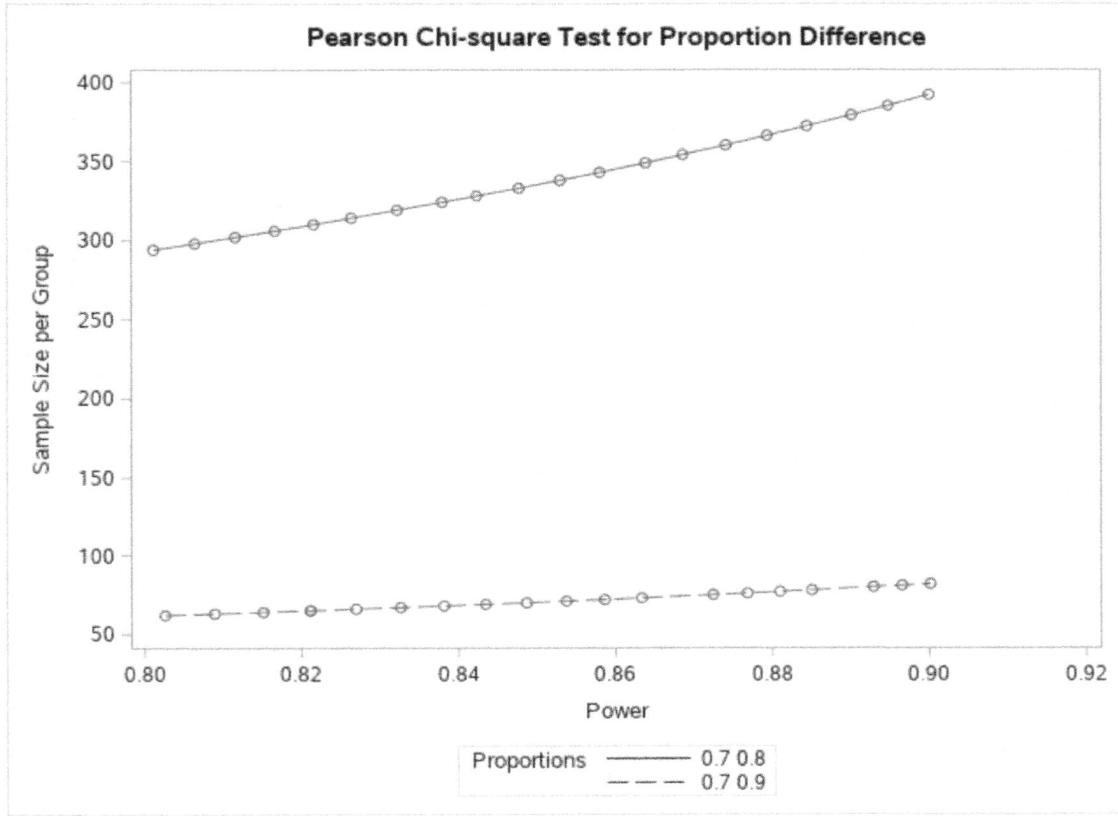

Figure 15.17: Graph of Results

Computing Sample Size for a One-Way ANOVA Design

The last example in this chapter describes how to compute sample size or power for a one-way analysis of variance (ANOVA).

In this example, you have made the following decisions:

- There are 3 means estimated to be 20, 25, and 30.
- You have two estimates for standard deviation: 8 and 10.
- You want to compute sample size for powers of 80% and 90%.
- You want to compute the n-per-group (as compared to power for a given sample size).

You would like a plot of Power (X axis) versus sample size, and the axes scaled to show powers from .7 to .9.

Start out by selecting Power and Sample Size for One-Way ANOVA. On the PROPERTIES tab, set the number of groups to 3, select Sample Size per Group, and then enter your hypothesized means, one or more estimates of standard deviation, and one or more values for power. (See Figure 15.18.)

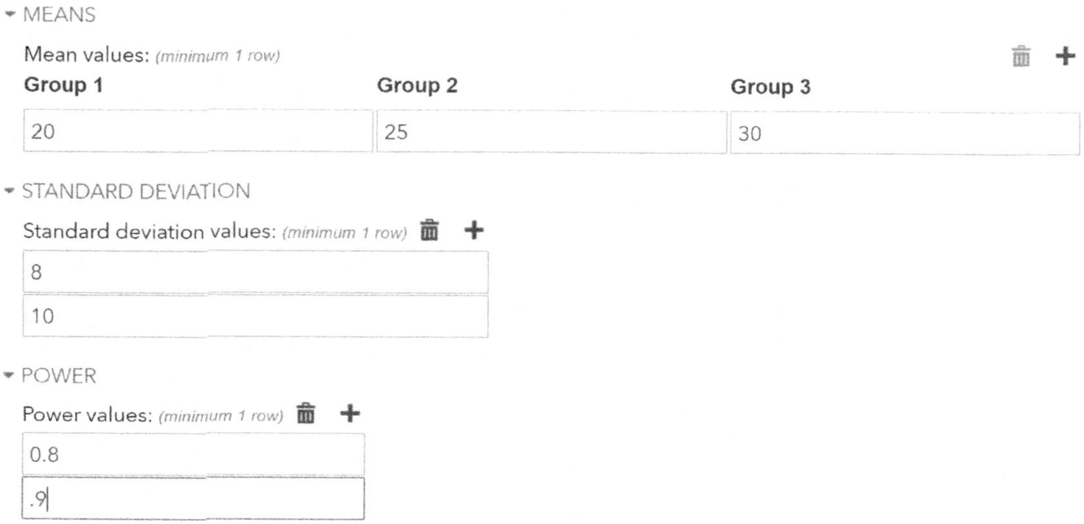

Figure 15.18: Entering Group Means, Standard Deviations, and Powers

The next step is to click the PLOTS tab. Select a plot of Power by sample size and enter a value for the minimum and maximum power (Figure 15.19).

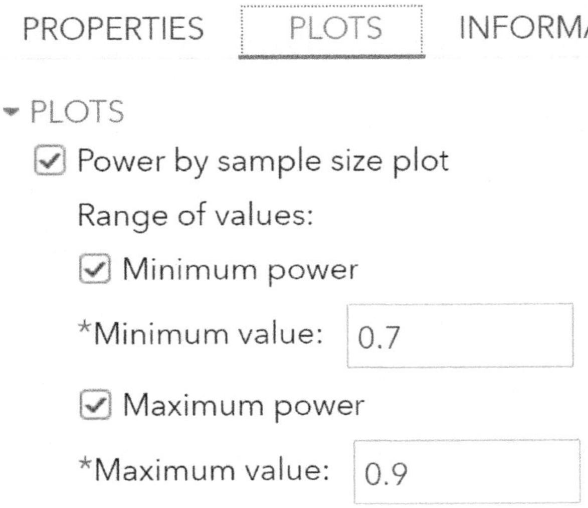

PROPERTIES PLOTS INFORMATION

▾ PLOTS

☑ Power by sample size plot

Range of values:

☑ Minimum power

*Minimum value: 0.7

☑ Maximum power

*Maximum value: 0.9

Figure 15.19: Selecting Desired Plots

You are ready to run the task. The first portion of the output is a table showing the required sample size for all combinations of estimated standard deviation and power. Because you entered two values for standard deviation and two powers, the table of sample size per group shows four different (2 times 2) values for sample size per group.

Computed N per Group				
Index	**Std Dev**	**Nominal Power**	**Actual Power**	**N per Group**
1	8	0.8	0.820	14
2	8	0.9	0.913	18
3	10	0.8	0.815	21
4	10	0.9	0.908	27

Figure 15.20: Table of N per Group

The plot request produced the graph in Figure 15.21.

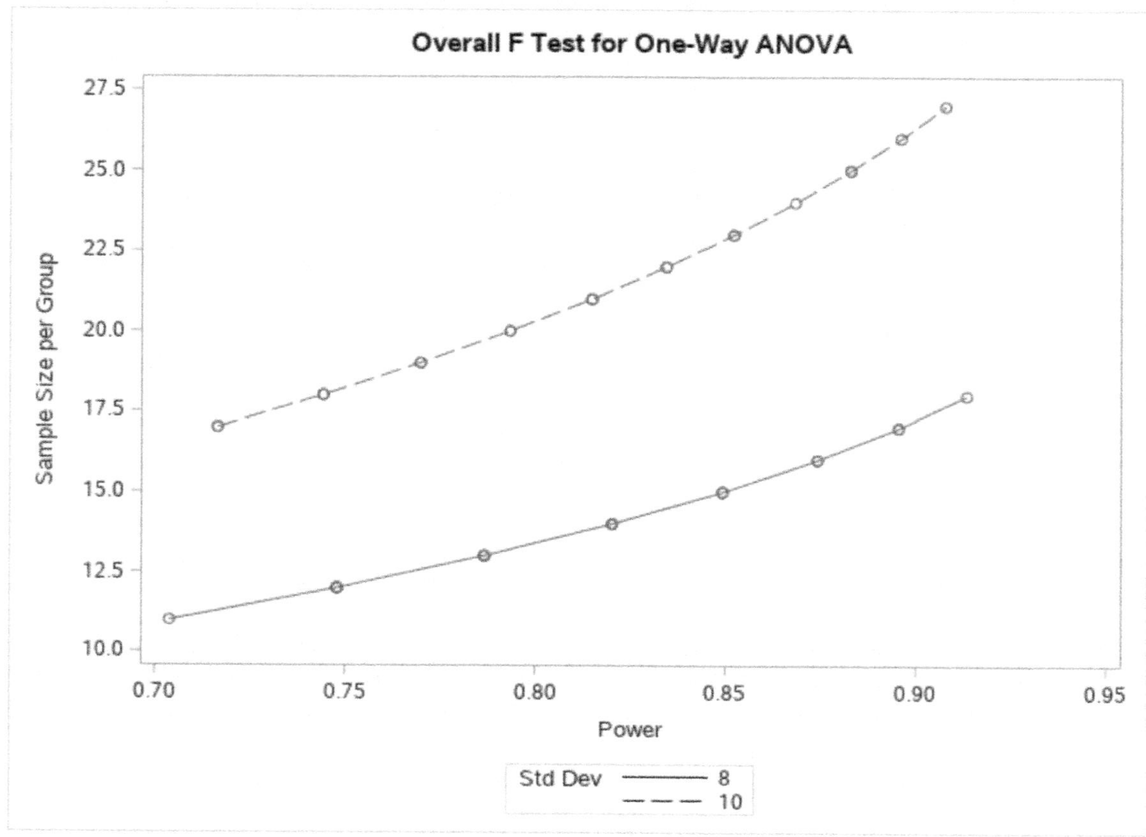

Figure 15.21: Plot of Power versus Sample Size

Inspection of this graph, and the practical constraints of obtaining and possibly paying for subjects, will help you make a final decision on designing your study.

Conclusion

The SAS Studio statistics tasks include power and sample size calculations for several statistical tests, such as comparing means, comparing proportions, or running an analysis of variance. By seeing how to enter data for the three scenarios in this chapter, you should be able to perform calculations for the other designs in the Power and Sample size menu.

Chapter 15 Exercises

1. Compute the sample size for a two-sample *t* test where you estimate the mean of the two

groups to be 200 and 210, respectively. You estimate the standard deviation to be 10, and you want to compute the number of subjects per group for powers of .80, .85, and .9.

2. The proportion of patients cured in under five days without any medication is .80. You want to compute sample size per group under two possible hypotheses. One with a five-day cure rate of .85, the other with a five-day cure rate of .90. Compute the sample sizes for powers of .8 and .9. Also produce a graph showing power on the X axis and sample size on the Y axis. Notice the huge difference in the required sample sizes for the comparison of .8 and .85 versus .8 and .9. In general, you need very large samples to test for small differences in proportions.

3. You are planning a one-way ANOVA design. You have two scenarios that you want to test. In one, the three hypothesized means are 50, 55, and 60. In the second, they are 50, 60, and 70. You also have two estimates of standard deviation: 15 and 20. Compute sample size per group for a power of .8. Include a plot of power versus sample size with the minimum power set at .7 and the maximum power set at .9.

16.

Solutions to the Odd-Numbered Exercises

Chapter 6 Solutions

1. Tasks ⇒Statistics ⇒Summary Statistics

   On the DATA tab, select the data set Iris from the SASHELP library.

   Under Analysis variables, select PetalLength and Petal Width and click OK.

   On the OPTIONS tab, check boxes for Mean, Standard deviation, Median, Number of Observations, Number of missing values.

   Under Additional Statistics, check Confidence limits for the mean.

   Under Plots, select Histogram.

3. Tasks ⇒ Statistics ⇒ One-Way Frequencies

   On the DATA tab, select the data set Iris from the SASHELP library.

   Under Analysis variables, select Species.

5. Tasks ⇒ Statistics⇒ Distribution Analysis

   On the DATA tab, select the data set Heart in the SASHELP library.

   Under Analysis variables, select Weight.

   Under Additional Roles and Group analysis by, select Sex.

   On the OPTIONS tab, check boxes for Normality and Histogram and Goodness of fit.

   The p-value for the Anderson-Darling test is <.005 to reject the null hypothesis that the data values are normally distributed.

7. Tasks ⇒ Graphs⇒ Box plot

On the DATA tab, select Cars in the SASHELP library.

Click Horizontal plot.

Under Analysis variables, select Invoice.

Under Category, select Cylinders.

Run this task and then go back to the DATA tab and add the filter: Cylinders=4 or Cylinders=6.

Chapter 7 Solutions

1. Tasks ⇒ Statistics ⇒ t-Tests

On the DATA tab, select the Heart data set in the SASHELP library.

Be sure One-Sample t test is displayed in the box.

Under Analysis variables, select Weight.

On the OPTIONS tab, be sure a two-tailed test is selected (it is the default).

The alternative hypothesis is mu ^= 150.

Check the box for a test of normality.

Under Plots, select histogram and box plots.

You should NOT be concerned that the tests for normality are significant. First of all, the histogram look symmetrical and the sample size is 5,000. With large sample sizes, tests for normality are often significant.

3. a) Tasks ⇒ Data ⇒ List Data

On the DATA tab, select the Air data set in the SASHELP library.

Under the List variables box, select Date and Air (the only variables in the data set).

On the OPTIONS tab, under Rows to list, select First n rows.

Enter 10 in the Amount box.

b) Tasks ⇒ Statistics ⇒ Summary Statistics

On the DATA tab, be sure the Air data set is still selected.

Under Analysis variables, select Air.

On the OPTIONS tab, select Plots and Histogram and box plot.

c) Tasks ⇒ Statistics ⇒ t Tests

On the DATA tab, make sure that a one-sample t Test is selected.

Under Analysis Variables, select Air.

On the OPTIONS tab, set the alternative hypothesis to mu ^= 285.

Check tests for normality and Nonparametric tests (Sign test and Wilcoxon signed rank test).

All tests fail to reject the null hypothesis. With an n of 144 and only a moderate tail on the distribution, you could probably use the t Test, but it would also be a good idea to run the Wilcoxon signed rank test as well.

5. Rerun the program with 20 changed to 200 and run rerun the task in exercise 4.

   All the tests for normality reject the null hypothesis. This is because you not have more power with a larger sample size. However, because this is a uniform distribution and n = 200, you can be comfortable running a t Test.

Chapter 8 Solutions

1. Tasks ⇒ Statistics ⇒ Tasks ⇒ t-Tests

   On the DATA tab, select the Heart data set in the SASHELP library.

   Select a two-sample t Test, Systolic Blood Pressure as your Analysis variable and Sex as your Groups variable.

   On the OPTIONS tab, be sure the box for Tests of normality is checked.

   You reject the null hypothesis of normality for both males and females, but with an n of 5,000 and a reasonable distribution, it is OK to run a t Test. Neither of the two p-values for the t Test are not significant.

3. You will find the file Ttest_Data.xlsx in the folder Statistics Data. If not, type in the data in Excel.

   Tasks ⇒ Utilities ⇒ Import Data

   Scroll down and click the Change box and name your file Ttest_Data and leave it in the WORK library.

5. Tasks ⇒ Statistics ⇒ t Tests

   On the DATA tab, select Ttest_Data in the WORK library and select a two-sample test.

   Select Score1 as the Analysis variable and Method as the Groups variable. The variances are not significantly different so that you can use the Pooled values. Score1 is significant at the .05 level.

   Tasks ⇒ Statistics ⇒ t Tests

   On the DATA tab, select the Cars data set in the SASHELP library.

   Create a filter that reads: Cylinders = 4 or Cylinders = 6.

   On the DATA tab, select Two-sample t Test.

   Select Invoice as the Analysis variable and Cylinders as the Groups variable.

   On the OPTIONS tab, select Tests for normality and the Wilcoxon rank sum test.

   Leave the defaults on the PLOTS tab.

   All statistical tests are highly significant.

7. Change the value on the DO loop to 50.

   This data set will be created when you run Create_Datasets.sas. You can also get it from the download (it's part of the Create_Datasets.sas program) or you can type it in.

   Tasks ⇒ Statistics ⇒ t Tests

   On the DATA tab, find data set TTest in the WORK library. Select a two-sample t Test.

   Select variable X as the Analysis variable and Group as the Groups variable.

   The p-value is .0094 (significant).

Chapter 9 Solutions

1. Tasks ⇒ Linear Models ⇒ One-Way ANOVA

   On the DATA tab, select the High_School data set in the STATS library.

   Select Vocab_Score as the Dependent variable and Grade as the Categorical variable.

   On the OPTIONS tab, make sure that the Tukey multiple comparison test is checked.

Grade is highly significant. Pairwise comparisons that are not significant are Junior vs Senior and Senior and Sophomore (close but > .05). Levene's test is not significant.

3. Run the program shown. That program is also in Create_Datasets.sas if you don't want to type it.

   Tasks ⇒ Linear Models ⇒ One-Way ANOVA

   On the DATA tab, select data set Temp in the WORK library, select Weekly_Salary as your dependent variable and Gender_Age as your Categorical variable. The ANOVA is highly significant as are all the pairwise comparisons.

5. Tasks ⇒ Linear Models ⇒ Nonparametric One-Way ANOVA

   On the DATA tab, select the Heart data set in the SASHELP library

   Enter Cholesterol as the Dependent variable and DeathCause as the Categorical variable.

   On the OPTIONS tab, select multiple comparisons.

   Pairwise differences less than .05 are: Other vs Coronary Heart Disease; Cancer vs Coronary Heart Disease; Cerebral Vascular Disease vs Coronary Heart Disease.

   Parametric analysis gives identical results.

7. Tasks ⇒ Linear Models ⇒ One-Way ANOVA

   On the DATA tab, select the Cars data set in the SASHELP library.

   Enter the following in the filter box: Cylinders in (4,6,8).

   Select Weight as the Dependent variable and Cylinders as the Categorical variable.

   All pairwise differences are highly significant.

Chapter 10 Solutions

1. Tasks ⇒ Linear Models ⇒ N-Way ANOVA

   On the DATA tab, select the High_School data set in the STATS library.

   Enter Vocab_Score as the Dependent variable, Gender and Grade as Factors.

   On the MODEL tab, click Edit, select Gender and Grade in the Variables list then click Full Factorial and scroll down to OK.

The interaction term is not significant and both main effects are significant at the .05 level.

The only pairwise comparison for Grade that is not significant is between Junior and Senior.

3. Tasks ⇒ Linear Models ⇒ N-Way ANOVA

On the DATA tab, select the Interact data set in the STATS library, select Parts as the Dependent variable, Training and Seniority as Factors.

On the MODEL tab, select Training and Seniority as Variables, click Full Factorial and OK.

Notice the highly significant interactive term. Go back to the OPTIONS tab. Under Statistics, select All effects in the pull-down list under multiple comparisons. Rerun the program. Notice that training has an effect on beginning workers but not the long-time workers.

5. Tasks ⇒ Linear Models ⇒ N-Way ANOVA

On the DATA tab, select the data set Salary in the STATS library.

On the DATA tab, select Weekly_Salary as the Dependent variable, Age_Group, Education, and Gender as Factors.

On the MODEL tab, click Edit, select Age_Group, Education, and Gender as Variables, click N-Way Factorial, select two-way interactions and click OK.

On the OPTIONS tab, select Plots and Suppress Plots.

Chapter 11 Solutions

1. Tasks ⇒ Statistics ⇒ Correlation Analysis

On the DATA tab, select the data set High_School in the STATS library.

Select the three variables as Analysis variables.

On the OPTIONS tab, select a Matrix of Plots for Type of plot and check the box for including histograms on the diagonal.

3. Tasks ⇒ Statistics ⇒ Correlation Analysis

On the DATA tab, select the Heart data set in the SASHELP library. Select Height, Weight, Systolic blood pressure, and Diastolic blood pressure as Analysis Variables.

On the OPTIONS tab, make sure the selection of None is in the box under Plots.

When you run it again with a request for individual plots, you do not get any plots. You need to increase the Maximum number of points to "No limit."

5. Tasks ⇒ Statistics ⇒ Correlation Analysis

   On the DATA tab, select the High_School data set in the STATS library.

   Select the variables Vocab_Score, Spelling_Score, and English_Grade in the Analysis Variables box. Select the variable Honor in the Correlate With box.

   On the OPTIONS tab, under Plots, select None.

7. Tasks ⇒ Statistics ⇒ Correlation Analysis

   On the DATA tab, select Physics_Test in the STATS library. Select the variables Ans1 through Ans10 for the Analysis Variables and Grade as the Correlate With variable.

   On the OPTIONS tab, under Plots, select None.

Chapter 12 Solutions

1. Tasks ⇒ Linear Models ⇒ Linear Regression

   On the DATA tab, select the data set High_School in the STATS library. Select Spelling_Score as the Dependent variable and Vocab_Score as the Continuous variable.

   On the MODEL tab, click Edit and select Vocab_Score in the variable box and then click Add. Finally, scroll down and click OK.

   The regression equation is Spelling_Score = 45.54663 + .74910×100 = 120.46

3. Tasks ⇒ Linear Models ⇒ Linear Regression

   On the DATA tab, select the High_School data set in the STATS library. Select English_Grade as the Dependent variable and Vocab_Score as a Continuous variable. In the box for Categorical variables, add Honor (0 or 1) and Gender.

   On the MODEL tab, click Edit and choose all four variables in the variable box. Then click Add and OK.

   On the MODELS tab, select Default and additional statistics. Under collinearity, check the box for Variance Inflation Factor. Even though none of the VIFs are high, notice that Vocab_Score is no longer significant and its beta (slope) is negative.

5. Tasks ⇒ Linear Models ⇒ Linear Regression

On the DATA tab, select the Cars data set in the SASHELP library, MSRP (manufactures suggested retail price) as the dependent variable and the three variables Horsepower, Weight, and Length as Continuous variables.

On the MODEL tab, click Edit, select the three predictor variables Horsepower, Weight, and Length in the Variables box, click Add and OK.

On the SELECTION tab, chose Stepwise selection and run the model. Only Horsepower and Length enter. Interesting to note that the coefficient for Length is negative. That's probably because of very expensive sports cars.

Chapter 13 Solutions

1. Tasks ⇒ Statistics ⇒ Summary Statistics

On the DATA tab, select the Graduate data set in the STATS library. Select English_Grade and Math_grade as Analysis variables, Study and Graduate as Classification variables.

If you want, you can open the OPTIONS tab and select comparative box plots and/or histograms.

3. Tasks ⇒ Linear Models⇒ Binary Logistic Regression

On the DATA tab, select the Graduate data set in the STATS library. Choose Graduate as the Analysis variable and 1 (yes) as the Event of interest. Select Study as a Classification variable and the two variables English_Grade and Math_Grade as Continuous variables. Select Reference coding as the parameterization method. Yes, the model improved as shown by a smaller value of SC (Schwarz Criteria), increased concordant pairs and fewer discordant pairs. If you choose, you can edit the code to modify the CLASS statement to read:

```
class Study (ref='0') / param=ref;
```

5. Tasks ⇒ Linear Models ⇒ Binary Logistic Regression

On the DATA tab, select the Graduate data set in the STATS library. Choose Graduate as the Analysis variable and 1 (yes) as the Event of interest. Select Study as a Classification variable and the two variables English_Grade and Math_Grade as Continuous variables. Select Reference coding as the parameterization method.

On the SELECTION tab, select Stepwise as the Selection method. If you wish, you can change the reference level for study to '0'. All variables are selected. Your edited code should look like this:

```
proc logistic data=STATS.GRADUATE;
```

```
class Study (ref='0')/ param=ref;

model Graduate(event='1')=Study English_Grade Math_Grade / link=logit

selection=stepwise slentry=0.05 slstay=0.05 hierarchy=single technique=fisher;

run;
```

Chapter 14 Solutions

1. Tasks ⇒ Statistics ⇒ One-way Frequencies

   On the DATA tab, select the Graduate data set in the STATS library. Under Analysis variables, select Gender, Study, and Graduate.

   On the OPTIONS tab, deselect Percentages and Cumulative frequencies and percentages. Under Plots, click Suppress all plots.

3. Tasks ⇒ Statistics ⇒ Table analysis

   On the DATA tab, select the Graduate data set in the STATS library. Select Study in the Row variables box and Graduate in the Column variables box.

   On the OPTIONS tab, be sure Chi-square is checked on the Statistics menu.

   Click CODE and Edit so that you can write your format. Your final program should look like this:

```
proc format;

value Graduate 1='1:Yes' 0='2:No';

run;

proc freq data=STATS.GRADUATE order=formatted;

format Graduate Graduate.;

tables  (Study) *(Graduate) / chisq nopercent norow nocol nocum plots=none;

run;
```

5. Tasks ⇒ Statistics ⇒ One-way Frequencies

   On the DATA tab, select the Fish data set in the SASHELP library. Select Species as the Analysis variables box.

On the OPTIONS tab, under Row Value Order, select Decreasing frequency. Make your own choices about cumulative statistics and plots.

Chapter 15 Solutions

1. Tasks ⇒ Power and Sample Size ⇒ t Tests

On the PROPERTIES tab, select Two-sample test under type of test. You want to solve for sample size per group. Leave the default selection of Pooled for Type of test. Under Select a Form, select Group means. Enter 200 for Group 1 and 210 for Group 2. Enter 10 as the Standard deviation. Under Power, enter .8, click the + sign, enter .85, click the + sign and enter .9.

3. Tasks ⇒ Power and Sample Size ⇒ One-way ANOVA

On the PROPERTIES tab, select Sample size per group. For the number of groups, select 3. Under means, enter the three means 50, 55, and 60, click the + sign and enter three values 50, 60, and 70.

Under Standard deviation, enter the two values 15 and 20 in the usual way. Enter a power of .8.

On the PLOTS tab, select Power by sample size and enter .7 as the minimum value and .9 as the maximum power.

Made in United States
North Haven, CT
25 January 2023

31619340R00167